The
WOOD BOOK

The
WOOD BOOK

An Entertaining, Interesting, and
Even Useful Compendium of
Facts, Notions, Opinions and Sentiments
about Wood . . .
and Its Growing, Cutting,
Working, and Burning

by Jan Adkins

Little, Brown and Company   Boston / Toronto

First Edition

Library of Congress Cataloging in Publication Data

Adkins, Jan.
  The wood book.

  Includes bibliographical references.
  1. Wood. I. Title.
TA419.A3   634.9′83   80-16739
ISBN 0-316-01082-0 (pbk.)

MV

Published simultaneously in Canada
by Little, Brown & Company (Canada) Limited

Printed in the United States of America

Photograph credits

The photographs appearing on pages 36, 37 courtesy of David Boye Knives; on page 52 courtesy of Segen Pax; on pages 70–71, 73–74, 82–83, 86–87 photographs by Benjamin Mendelowitz, courtesy of *WoodenBoat*; on page 72 photograph by Peter Leth, courtesy of *WoodenBoat*; on page 76 photograph by Douglas Photo Shop, courtesy of *WoodenBoat*; on page 78 courtesy of *WoodenBoat*; on pages 93, 94, 95, 96, 98 courtesy of Carol Link; on pages 115, 119 courtesy of Conover Wood Craft Specialties, Inc.; on pages 135, 136, 137, 138 courtesy of Dean Torges; on page 148 courtesy of Baldwin Tubs; on page 151 courtesy of Suzanne Page; on pages 157, 158, 159, 160 courtesy of *CoEvolution Quarterly*; on page 176 courtesy of The Renwick Gallery, National Collection of Fine Arts, Smithsonian Institution, Washington, D.C.

The
WOOD BOOK

was so young that I sat on his shoulders and the lowest leaves brushed my face with the same light, dry caress my great-grandmother gave. "Say goodbye," my father said. I said goodbye and waved and then we were past the clearing, the talking from the porch behind us, only a hoot of laughter slipping to us through the trees, then only the trees. High up on his shoulders the tunnel of road ran on through shadow and sunlight and dark and bright until

A Wood List

by Jan Adkins

I can never dawdle through a list of woods without a fit of wanderlust and the lumberjack fantasy: the Great Deep Green Forest, stagged wool pants as thick as deck canvas, and the sputtering hiss of Paul Bunyan's bull cooks skating on a half-acre flapjack grill with a sow-belly strapped to each foot. Timber! Each wood announces its strengths, reveals its vulnerabilities, and cites the marvelous products that take advantage of its qualities. With a good list, I think, a really good list, I could make a chair or a cockpit grating or a chest of woods so appropriate to the purpose and congenial to each other's needs that they would last as long — no, longer — than that model of compatibility, the Deacon's Wonderful "One-Hoss Shay." Ah, well, I say, the Deacon had the whole countryside's pick of fine seasoned wood and what have I got?

In truth, the wood available to you is probably as good as or better than the wood used by the Deacon's contemporaries. Seventeen fifty-five (by the village clock) hadn't the network of communication and delivery at your service. Woods of every variety, from all over the world, are ready for you at a phone call and a check that would have bought the shay

outright. Though the wood is now second-growth stuff, at least you can have the best quality and not the year's remnants. We have other problems: more wood is available but little is known and less is passed on about the traditional and practical ways of working individual species. When lumber becomes so dear that its price affects your work, you stand at a disadvantage. High-priced, low-quality construction lumber discourages workmanship in the commonest odd jobs of occasional framing or set up, which were training grounds for more difficult pieces and often, in slower times, works of some ingenuity and polish. It's sad not to have more common species, more hardwoods and clear softwoods at a price. Hem-fir, that mythic tree, grows a poor stuff with which to work.

A wood list is most seductive in its "uses" section. Abstracting the reasons for specific purposes begins to fill in a composite picture of a wood. The lineage of its uses insists that you utilize a wood well, play to its strengths and finesse its weaknesses. Its wood-list properties suggest applications and encourage a revelation of character in the piece of work under your pencil: look at the strength in this

it bent and the forest ate it. The patches of sun squinted my eyes and the brightness made the lower wall of the tree tunnel opaque, but in the green shade the depth of the woods reappeared and the broken view of a parallel tunnel around the creek, which glinted and sparkled in its own pattern of light and shade.

The trees began high in sugar maples and oaks, hickories, walnuts, and reached across the road in sycamores. Oaks and willows and

thin shaft of hickory, see how this elm seat refuses to split. Wood lists and theory are late-night entertainment, and they are no more than smudged pencil notes beside the reality of the wood you finally find yourself alone with, in your shop, giving over the pencil designing to cutting edges. It is good planning and a base of design to suit wood to its use, and a wood list is a help; the final statement, however, should grow with the grain out of the wood itself.

There are no listings here of exotic woods, domestic or foreign, none of the rare and lovely stuff of marquetry and inlay and high-altitude joinery. There are very good books describing them and showing them in four-color process reviewed here. In this list are available American woods that may well be all you ever need, certainly enough quality and strength and color to show your most advanced skills. My two advisers differ, here: John Swain Carter maintains that local woods are close to you, part of your ethos, and you ought to stay within the provincial range of flora. The Dorothy-returning-to-Kansas theory is sound, and if you can't find your heart's desire far from home maybe you never lost it in the first place and maybe you haven't tried hard enough. Dean Torges, who uses mostly local woods, maintains that Dorothy found the truth of Kansas only by leaving for Oz, and that exploration is the deep source of inspiration. With a shrug and a click of our ruby slippers we present a partial, a fallible, a provincial and provisional list of woods.

Alder (red alder)   *Alnus rubra*

Moderately light and strong but poor shock resistance, low shrinkage, not a durable wood
Specific gravity = .46   app. 29 lb/ft³
Millwork, furniture, sash stock, doors, veneer, brush and broom backs, gunpowder charcoal

White Ash   *Fraxinus americana*

Strong and stiff, shock resistant, with excellent bending qualities
Specific gravity = .67   app. 42 lb/ft³
The wood of choice for shovel handles; for sports equipment such as bats, sculling oars, etc.; bent furniture and veneers

Aspen
(quaking aspen)   *Populus tremuloides*
(bigtooth aspen)   *Populus grandidentata*

Light, weak, soft, fairly stiff, breaks down quickly in weather
Specific gravity = .35–.43   app. 23–27 lb/ft³
Paper pulp, crates, boxes, excelsior, matches

Basswood (American   *Tilia americana*
                linden, lime)

Stable and easily worked, though not strong or decay resistant
Specific gravity = .42   app. 26 lb/ft³
Carvings, musical instruments, slack cooperage, beehive frames, piano parts, small turnings, veneer, chests, seats

Beech   *Fagus grandifolia*

Strong, works well, excellent for turning, takes a high polish, unstable, subject to decay, shrinks considerably in drying
Specific gravity = .72   app. 45 lb/ft³
Furniture, flooring, ties, barrels for food, veneer, spoons (no taint), brush handles, chairs

Beech

Paper Birch

Birch

(yellow birch)	*Betula allegheniensis*
(sweet birch)	*Betula lenta*
(paper birch)	*Betula papyrifera*

Tough, strong, turns very well, susceptible to decay, takes a good polish, too brittle for steam bending, splits near ends

Specific gravity (yellow) = .64   app. 40 lb/ft^3
(sweet) = .76   app. 48 lb/ft^3
(paper) = .59   app. 37 lb/ft^3

Yellow birch: furniture, veneer, flooring, treenware, wheel hubs

Sweet birch: furniture, millwork, treenware, boxes, pulp; distilled for alcohol; twigs and bark produce birch oil for birch beer and flavoring of medicines

Paper birch: clothespins, shoe lasts, turnery, spools, fuel, pulp

Buckeye

(horse chestnut)	*Aeschulus glabra*
(Ohio buckeye)	*Aeschulus hippocastanum*

Weak, soft, brittle and vulnerable to decay, it is nevertheless stable and light, finishes poorly, resists splitting, easily cut and carved

Specific gravity

(horse chestnut) = .56   app. 35 lb/ft^3
(Ohio buckeye) = .43   app. 27 lb/ft^3

Of little commercial importance it nevertheless has unusual specialty uses due to its resistance to splitting, weight, and stability once seasoned: engineering patterns and artificial limbs

Butternut (white walnut)   *Juglans cinerea*

Coarse, weak, soft, high in shock resistance, machines and finishes well

Specific gravity = .59   app. 37 lb/ft^3

Furniture, cabinets, trim, nuts

Cedar
(eastern red cedar)   *Juniperus virginiana*
(western red cedar)   *Thuja plicata*

Two cedars of good decay resistance, the eastern heavier and denser, both fairly low in strength, stable when seasoned, easy to work

Specific gravity   (eastern) = .51   app. 32 lb/ft³
                    (western) = .45   app. 28 lb/ft³

The western variety supplies about half of our shingle supply and is a coarser wood; the eastern takes a good finish and is used for chests, pencils, closet linings, cigar boxes, canoes, treenware, and for its aromatic oil

Alaska Cedar
(Alaska cypress) *Chamaecyparis nootkatensis*

Heavy, strong, stiff and hard, very resistant to decay, stable

Specific gravity = .50   app. 31 lb/ft³

Small boat building, canoe paddles, patterns, trim, furniture

Incense Cedar   *Libocedrus decurrens*

Light, weak, fragrant, decay resistant

Specific gravity = .40   app. 25 lb/ft³

Standard grades are used for posts and sash stock, better grades for closet linings, shingles, venetian blinds, and pencils

Port Oxford Cedar   *Chamaecyparis lawsoniana*

Fairly strong and hard but light and stable, fine-grained, easy to work, fairly resistant to shock, exceptionally decay resistant, takes a fine finish

Specific gravity = .46   app. 29 lb/ft³

One of the most durable woods, it is used for battery separators, outdoor construction, boatbuilding, arrow shafts, closet linings, and as a basis for sweet-smelling resin used in soaps and cosmetics

White Cedar
(northern white cedar)   *Thuja occidentalis*

Very light, comparatively weak, but stable and durable in conditions encouraging decay, splits easily and is soft but works well and takes a good finish

Specific gravity = .30   app. 19 lb/ft³

Because of its easy splitting this is a traditional wood for canoe ribs; its weather resistance makes it a specialty wood for tanks and some boatbuilding applications

Cherry (black cherry)   *Prunus serotina*

Strong, relatively light, works very well and takes a high finish, very resistant to shock, fairly stable when seasoned, darkens with age

Specific gravity = .57   app. 36 lb/ft³

One of the premier cabinetmaking woods for furniture, panels, fine joinery and veneer; also a staple for backing electrotype blocks

Chestnut
(American chestnut)   *Castanea dentata*

Soft, weak, but stable and easily worked, acidic and tends to corrode iron fastenings, highly decay resistant, easily split

Specific gravity = .48   app. 30 lb/ft³

Its decay resistance has made it a traditional choice for ties, fencing and fence posts, drawer bottoms, barn beams, caskets, and poles; also seen as veneer core stock and as "wormy chestnut" in trim and picture framing

Cottonwood
(eastern poplar)   *Populus deltoides*

Weak, soft, warping badly in seasoning, it is easy to work and light

Specific gravity = .38   app. 24 lb/ft³

A utility wood for cartons, boxes, pulp, fuel, and veneer core

sycamores stepped down to the creek and leaned over its summer slow progress. They marched past my mahout's perch undistinguished, wearing the repeated green face of the forest. Below their crowns the spindly hopefuls survived, now and again a quaver of fruit trees, gnarled beyond giving, haunting a long-gone homesite. Below them the bull briars, elderberries, the seedlings and then the perversely thorny and delicious black rasp-

6

American Elm

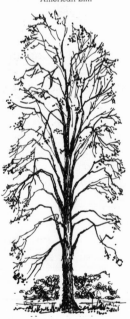

Shagbark Hickory

Bald Cypress   *Taxodium distichum*

A wood of medium weight and strength, it is among the most decay resistant, easily worked

Specific gravity = .56   app. 35 lb/ft³

Exposed construction, tanks, vats, shipbuilding, stadium seats, greenhouse equipment, caskets, cooperage, ties, posts

Elm (white elm)   *Ulmus americana*
　　　(rock elm)   *Ulmus thomasii*

Hard, heavy, tough, strong and very difficult to split, will survive indefinitely if constantly wet, works well but does not take a good polish

Specific gravity

(white elm) = .56   app. 35 lb/ft³

(rock elm) = .70   app. 44 lb/ft³

Chair seats, wheel hubs, bowls, and butcher blocks use elm's resistance to splitting, and its resistance to decay when in water makes it a wood traditionally used for coffins, shipbuilding, and water conduits

Fir (noble fir)   *Abies procera*
　　(white fir)   *Abies concolor*

Light, even-grained with an attractive contrast, soft, poor weather resistance

Specific gravity = .42   app. 26 lb/ft³

Standard grades go to house construction and plywood core and pulp; best grades make up interior finish, siding, sash, stair treads, door stock, venetian blinds and ladder rails; specially selected lots are used for airplane construction

Douglas Fir   *Pseudotsuga menziesii*

Varies widely in strength and properties according to growing conditions, high strength-to-weight ratio, resistant to decay, size makes knot-free pieces easily available

Specific gravity = .50   app. 31 lb/ft³

berry bracketed the dirt road at the clearings.

We were making a visit, a kind of yearly pilgrimage for us. It was made on foot, his feet for the most part, but I had my time on the ground, passing undignified flat frogs with interest and squatting beside dead snakes stopped and held at a moment of flight from the rumbling road with a part of their reptile intensity still in them, undusted.

The winters between those summers come

Of enormous commercial importance, the monoculture of Douglas fir has taken the western forests; used in construction, trim, millwork, and in every utility usage

Hemlock
(eastern hemlock)   *Tsuga canadensis*
(western hemlock)   *Tsuga heterophylla*

Light, soft, splintery, but moderately strong, poor weather resistance

Specific gravity   (eastern) = .45   app. 28 lb/ft³
                    (western) = .46   app. 29 lb/ft³

A utility lumber for construction use, framing, subflooring, boxes, pallets, concrete forms, etc.

Hickory (shagbark hickory)   *Carya ovata*

The strongest and toughest commercial wood, heavy and resistant to shock, stiff

Specific gravity = .76   app. 48 lb/ft³

Hickory's great strength and special stress properties make it the choice for hammer handles, all tool handles, wheel spokes and wagon axles, chair and stool legs, wagon shafts and singletrees, parallel bars; chips and waste are used to smoke food

American Holly (white holly)   *Ilex opaca*

Tough, close-grained, shock resistant but weak, takes a good polish, poor weather resistance

Specific gravity = .76   app. 47 lb/ft³

Cabinet inlay, turnery, musical and scientific instruments, piano keys, sometimes alternated with walnut as cabin sole in yachts, wood engravings, spoons and kitchen ware

Because of difficulties in drying, holly is usually had in small pieces

Hornbeam (ironwood)   *Carpinus caroliniana*

Tough, dense, hard, strong, unstable, hard to split, difficult to work but turns well and takes a good finish, not for outdoor use without preservative

Specific gravity = .78   app. 49 lb/ft³

Tool handles, windmill and watermill machine parts, parts for cider mills, plane bodies, billiard cues, drumsticks, piano parts

Larch (western larch)   *Larix occidentalis*

The heaviest of the conifers, this is a fairly strong wood with a tendency to split cleanly, one of the most durable woods in contact with the soil, it works well

Specific gravity = .61   app. 38 lb/ft³

Because of its resistance to ground rot, this construction timber is used for mine timbers, fence posts, ties, telephone poles, and boats, as well as in furniture and box construction.

Locust
(honeylocust)   *Gleditsia triacanthos*
(black locust)   *Robinia pseudoacacia*

Very heavy, very hard, very strong, stiff, shock resistant, stable after seasoning, difficult to machine, very decay resistant

Specific gravity
(honeylocust) = .70   app. 44 lb/ft³
(black locust) = .76   app. 48 lb/ft³

Fence posts, poles, mine timbers, and the insulator pins on telephone pole crossyards are locust because of its decay resistance; wooden ships are also fastened with locust treenails and many ladders have locust rungs

Soft Maple (red maple)   *Acer rubrum*

Strong, dense, hard, stiff, shock resistant, considerable shrinkage, turns well, poor weather resistance

Specific gravity = .49–.54   app. 34 lb/ft³

Furniture, cabinets, interior trim, gunstocks, veneers, treenware, flooring, tool handles

back as voids, blacks and cold grays, now: city winters, cold pavement and the sound of buses rattling dishes in the cupboard. Summers, though, are warm and colored, mostly tree green and the color of sun on that dirt road.

We walked toward Turkey Run but turned short of it and worked back and up the ridge on another road, climbing until I was on his shoulders again and could see the creek below us and the field beyond. Then at a switchback

PLANTING TREES

"We have all been sitting under trees planted by our ancestors, so let us plant trees so that our children and grandchildren can sit under them and possibly get the harvest."
— E. F. Schumacher

A tree can be grown from seed or purchased from a nursery. Planting procedures vary greatly, but the following examples are enough to begin:

Growing black walnut from seed
1. Collect several black walnuts in the fall. Remove their husks.
2. Select the site for the future tree. Be aware that a mature tree can be fifty feet high and its roots will damage or kill apple trees and some vegetables.
3. Loosen soil two feet deep and two feet in diameter. Plant four nuts two inches deep about six inches apart. Mulch the soil with several inches of shredded leaves.
4. After planting, cover the spot with wire mesh (to obstruct squirrels). Remove the mesh in early spring.
5. When seedlings appear, carefully remove all but the best one. Mulch the young tree to suppress weeds.

Growing apples from seed
1. Eat an apple in the fall and save the seeds.
2. Select a site for the future tree and loosen soil two feet deep and two feet in diameter.
3. Plant six apple seeds one inch deep, evenly spaced. Mulch with several inches of shredded leaves.
4. When seedlings appear in spring, carefully remove all but the best one. Let it grow through the summer, fall, winter, and spring until the next June. Mulch well.
5. In June, read up on bud grafting and bud onto your seedling a piece of the best, hardiest, most disease-resistant apple tree in your area.
6. Mulch the new tree to suppress weeds, wait for apples.

Planting potted fruit trees from a nursery
1. Dig a hole with vertical sides one foot larger in diameter and one foot deeper than the tree root-ball. Loosen the soil

in the bottom of the hole an additional eight inches and mix in some topsoil and old compost.
2. If the tree is wrapped in burlap, loosen the top and plant the whole ball (with burlap). If the tree is wrapped in plastic or is in a pot, cut away the container carefully.
3. Plant the root-ball with the plant at the same level to the ground as it was to the pot. Fill all spaces around the roots with good topsoil, pressing firmly to eliminate air pockets; when half-filled with soil, soak the hole with water to aid settling of soil.
4. Fill the hole with soil to ground level. Leave the top surface covered with loose earth and mulch to absorb rainfall more easily.

Planting bare-rooted fruit trees from a nursery
1. Keep trees cool, moist, and away from wind and sun until planting.
2. Dig a hole with vertical sides, one foot larger in diameter and one foot deeper than the root spread. Loosen the soil in the bottom of the hole an additional eight inches, and mix in some topsoil and old compost.
3. Set the tree two inches deeper than it stood in the nursery. Fill in around the spread roots with good topsoil, adding water to settle the soil.
4. Leave the soil surface loose and slightly dished to absorb rainfall. Mulch with hay, strawy manure, or peat.

For further details, find in your library: H. T. Hartmann and D. F. Kester, *Plant Propagation: Principles and Practices*, 3rd edition (Englewood Cliffs: Prentice-Hall, 1975).

Earl Barnhart is a tree man, a careful Johnny Appleseed who knows the numbers and books of leaves and bark and roots. He loves his leafy friends and is at the New Alchemy Institute, a coven of biological wizards in Falmouth, on Cape Cod.

turn the object of our pilgrimage: carved in the bole of what must have been an oak was the fantastic image of a bull's head, using the whorls and folds of the tree flesh as angry features out of a fevered dream or a Siamese temple, a reality more intense for the rude directness of the hand that carved it. What fascinated me about it, still fascinates me? Far from any home, carving it was an obsessive act suggested by half an image hidden in the

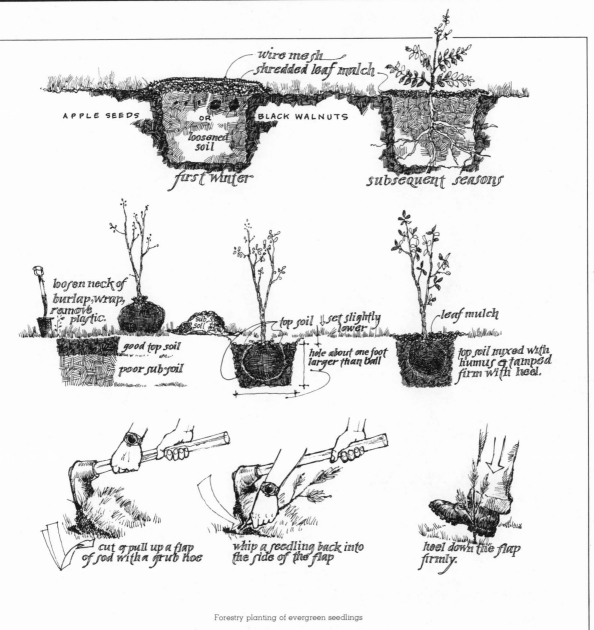

Forestry planting of evergreen seedlings

tree itself, completed by a man who had the other half in a mind that knew the power and potency and danger and precipitous unreason of bulls. How had the tree called to that man? He stood on the side of this hill for many hours joining his half of the image to the tree's. What possessed him? Far from any human structure, save the road, the Bull Tree had become a local deity, a small icon fixed in the tree, such as call to travelers all over the world in a thou-

Sugar Maple

Red Oak

Hard Maple
(sugar maple, rock maple)   *Acer saccharum*

Strong, very hard, stiff, strong, shock resistant, close-grained, takes a good polish, considerable shrinkage, turns well, poor weather resistance
Specific gravity = .68   app. 44 lb/ft^3

Furniture, flooring, gunstocks, bowling pins, musical instruments, cabinets, tool handles

Maple sugar — each tree of average size can produce almost three pounds of sugar

Accidental figures highly prized: bird's-eye, curly, flame

Red Oak
(northern red oak)   *Quercus borealis*
(swamp red oak)   *Quercus falcata*
                            (var. *pagodaefolia*)

Heavy, fairly strong, hard, more difficult to machine and work than white oak, tendency to split, hasn't the weather-resistant qualities of white oak, acid nature tends to corrode iron fastenings
Specific gravity
(northern red) = .72   app. 45 lb/ft^3
(swamp red) = .76   app. 48 lb/ft^3

Interior paneling, flooring, furniture, construction timbers

White Oak
(white oak)   *Quercus alba*
(chestnut oak)   *Quercus prinus*
(post oak)   *Quercus stellata*
(California white oak)   *Quercus lobata*

Hard, heavy, tough, stable, difficult to work, a regular grain whose structure makes it impervious to water, moderately decay resistant, a tendency to split
Specific gravity
(white oak) = .76   app. 48 lb/ft^3

sand species of roadside oaks and baobabs and yews and nux vomica. A small god of local powers with shade and roots, and I remember it each summer.

In the nation of trees every community has a patriarch, a tree notable for its age or girth or height or its great sky-covering crown, even its hunched and twisted limbs. It may be part of a center of a community, the Settler's Elm, the Charter Oak, the Fathers' Tree; or it

(chestnut oak) = .75   app. 47 lb/ft³
(post oak) = .75   app. 47 lb/ft³
(California white) = .64   app. 40 lb/ft³
Construction timber, cooperage, flooring, furniture, paneling, pilings, veneers, shipbuilding, ties, mine timbers, truck floors

Pecan (pecan hickory)   *Carya illinoensis*
Hard and heavy with many of the same strengths and working properties as hickory
Specific gravity = .73   app. 46 lb/ft³
Furniture, flooring, handles, fuel, nuts

Pine
(eastern white pine)   *Pinus strobus*
(sugar pine)   *Pinus lambertiana*
These two species make up the bulk of pine not used for construction; they are both fairly strong, light woods with even textures, ease of workability, and high stability
Specific gravity
(eastern white) = .44   app. 28 lb/ft³
(sugar) = .46   app. 29 lb/ft³
Stability makes both good woods for pattern making and furniture, paneling, sash stock, etc.

Pine (southern longleaf pine) *Pinus palustris*
Extremely strong, very hard, remarkably durable, finishes well and is resistant to decay
Specific gravity = .67   app. 42 lb/ft³
Because of its strength and the availability of large sizes, this has been the choice for construction timber, shipbuilding timber, and mine timber, flooring, masts, spars, ties, and also for its distilled sap in the form of turpentine, pitch, and resins grouped under the title of "naval stores"

Sassafras   *Sassafras albidum*
Fairly heavy and hard, weak in bending and end

grain compression, shock resistant, resistant to decay
Specific gravity = .51   app. 32 lb/ft³
Weather-resistant characteristics make this a choice for small boat building, foundation posts, sills for house construction, fences, and exposed trim; best known for its elixir from roots and bark, sassafras tea

Sweetgum (redgum) *Liquidambar styraciflua*
A strong, moderately heavy wood with an interlocking grain and a red-brown luster and figure that make it one of the most beautiful American woods
Specific gravity = .58   app. 36 lb/ft³
Furniture, trim, veneer, cigar boxes, baskets, ties

Sycamore
(plane, buttonball)   *Platanus occidentalis*
A difficult interlocking grain makes the wood hard to split or work, fairly strong with an excellent shock resistance, often has a pleasing figure when quarter sawn, easily decayed
Specific gravity = .57   app. 36 lb/ft³
Butcher blocks, saddletrees, musical instruments, cigar boxes, veneer, slack cooperage, handles

Tuliptree
(yellow poplar)   *Liriodendron tulipifera*
Light, soft, easily worked, takes a good paint finish, stable after seasoning
Specific gravity = .42   app. 26 lb/ft³
Furniture, veneer, musical instruments, piano cases, slack cooperage

Tupelo (blackgum, sour-gum) *Nyssa sylvatica*
Soft, weak, but resistant to splitting, tough, poor weather resistance
Specific gravity = .56   app. 35 lb/ft³

may be apart and away in a forest of lesser retainers, or solitary like the landmark trees that stood alone on the prairie. They are often benevolent and graceful. Some are places of haunts and visions. Some are hanging trees. They have powers. With their endurance and age they bind our history across the brief spans of generations. Their mass, their presence, centers our regard for the whole voiceless

Furniture, molding, floors, ox yokes, hatter's blocks, glass rollers, gunstocks, pistol grips

Walnut (black walnut)   *Juglans nigra*

Strong, uniform, shock resistant, weather resistant, stable after seasoning, easy to work, takes high finish
Specific gravity = .62   app. 39 lb/ft³
Cabinets, furniture, veneer, paneling, gunstocks, bowls, trim, delightful nuts

Willow
(black willow, swamp willow)   *Salix nigra*

Weak, soft, light, withstands splintering, poor weather resistance
Specific gravity = .42   app. 26 lb/ft³
Artificial limbs, cricket bats, and polo balls are made of willow because of its ability to withstand shock without splintering, malt house floors also use this property; to some extent fire-resistant, it is also used on brake blocks for machinery, and since it will last indefinitely if constantly wet (like elm) it was used for steamboat paddle blades and for underwater bulkheads along rivers

Redwood   *Sequoia sempervirens*

Light and fairly strong, easy to work, takes a good finish, resists decay, soft, stable
Specific gravity = .40   app. 25 lb/ft³
Its decay resistance and the availability of large sizes make this ideal for bridge timbers, silos, tanks, vats, ties, outdoor construction, shingles, caskets, furniture, and many other uses, but the scarcity of the great trees makes their use questionable

Spruce
(eastern spruce)   *Picea rubens* (red)
                     Picea glauca (white)
                     Picea mariana (black)
(Engelmann spruce)   *Picea engelmannii*
(Sitka spruce)   *Picea sitchensis*

Fine, straight-grained wood with good stability
Specific gravity = .44   app. 28 lb/ft³
The eastern and Engelmann spruces are harvested largely for pulpwood, but they are also used for framing lumber, piano sounding boards, and railroad ties; Sitka spruce, in its highest grades, is used for airplane construction and musical instruments, ladder rails, furniture, and lengths of straight-grained stuff are prized for masts and spars

kingdom of plants, so pervasive, so powerful in its passive dominance of land and sea.

Frail human belief wants to reside in something lasting and strong, a local deity of dusty switchbacks or the mighty Yggdrasil: for the Norsemen it held the sky in its branches, supported the earth with its trunk, and sank its three great roots into the underworld, one evergreen ash that held earth and

Wood Grades

by Jan Adkins

A professional cabinetmaker stocks his hoard of timber with the care and foresight of a sommelier, laying down a few pieces or whole trees against distant needs. Lodged in the back of his memory is an index of what is ready, and he will save an especially vibrant block of zebrawood for a special project as a sommelier might retain a bottle of promising Trockenbeerenauslese for the flourish at the end of a celebration. For almost any job his woodpile has a stick or a plank to answer.

The occasional woodworker doesn't have the advantage of a woodpile. He is forced to buy his raw materials from a mill or a lumber dealer and part of the creative contribution of the wood is lost to him. He can't live with a plank of cherry a year or two, letting it suggest uses and forms to him; he goes to fetch his wood and brings back a stranger. He faces frustration when he determines to cobble up a walnut chair and finds that, if his mill has any walnut at all, it is too thin, too narrow, too short, or full of shakes and wormy. At best he can approach the lumber shed with plenty of time and a flexible spirit and an option to say "The hell with it for now."

Even if the project is as pedestrian as a fence or a potting shelf, frustration waits at the lumberyard in the stacks of wet, warped 2×4s stored in the open, in the splintered, disappointingly rough plywood, in the raw quality of available wood.

In almost any area, though, it ought to be possible to find a good lumber source and to find a workable piece of stock worthy of your efforts, especially if you temper your needs with a few cautions and tools.

Caution your sense of design not to be too rigid. Use a wood that suits the purpose but bend toward local woods. This is not only good sense but good design. A concept should grow out of its time and its place as well as its designer. Some of the most charming pieces of woodwork, like Appalachian dulcimers or New England fine chests, have the strength of identity; they are part of their own forests. Utilizing what is available is an exercise in restricted palette and can produce some pleasant surprises. Richard Basins of Indian Ridge Traders (dealers in knife blades and knife-making supplies) reports that noncommercial woods, especially the woody shrubs and thorns, make beautiful knife and tool handles. Beyond convenience and design criteria, buy-

heaven and hell together, a tree of life, a tree of time and space and spirit. The trees of Knowledge and Life grew in the Garden of Eden and we ate from only one before that forest was lost to us. The burning bush, the sacred shittahs (acacia), a sprig of olive carried back to the gopher wood ark, cedars holy to Hebrew, Muslim, and Christian, the cherry

Alligator, or winching scow, kedging itself overland

ing local wood stimulates local lumbering and helps insure that hemfir will not supplant every other commercial wood.

Bill Gilkerson, a marine artist and piper, was researching the Highland bagpipes during a stay in Scotland. He asked an old piper what woods made up the turnings and the old man replied that, in those parts, the pipes were turned from oaks beyond a certain marsh and from larch, only a bit more distant, around the mountain. "I would have thought," Bill ventured, "that they would have had a lot of problems with moisture damage." "Ah, sure," the old man replied, "check and crack like the very devil. Right down the barrel. Awful."

"Well, then, they're not the best woods for the purpose."

"Na, na. They're terrible woods for pipes, but that's the tradition of it. That's what's always been used. That's what they had at hand, you see."

With modern transport lumber comes to us from all over the world and we can choose

wood that fits the requirements of the job at hand, but keep in mind that forests themselves vary naturally and that, just as there are a dozen species of tree referred to as "ironwood" in their own localities, there is probably a local species that can fit your requirements cozily.

When you go to the mill, caution yourself against rush. Try to let the wood speak to you. Think of the sizes, species, the markings and the grain as a dialogue between the wood and the designer. Take a tape measure and a board-foot chart (there is one on page 18) and a calculator and compare prices between species and pieces. Whether you are after 2-inch cherry or hemfir 2×4s, overbuy if your checkbook allows it; prices will only go up, and you can build a supply of dry, ready wood.

In all likelihood you will be sent to a drafty shed and set to picking over hardwood planks or told to back up to the stud pile and you will make do with what's there, or you won't. It may be useful at some time, though, to be familiar with lumber grades. At first it seems

that bent its branches down to Mary, the grim tree of crucifixion: if not holy themselves, like the Druid oaks, they can be caught up in holiness. The bo tree under which Buddha meditated until he found the way to Nirvana gave respect to all its family, like the banyan whose form Brahma took when he passed from this incarnation.

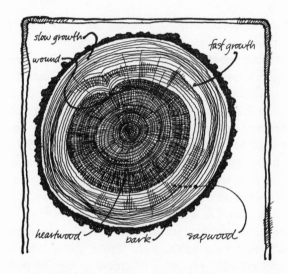

that lumber grading is purposely kept complex to confuse the wood buyer. It seems like that at the end, too.

For grading, lumber is separated into the two traditional categories which mean much more to the forester than to the woodworker. In these categories of "hardwood" and "softwood" are subcategories, each graded by different standards and different agencies. So much for simplicity.

Hardwoods fall into three groups according to their use. *Factory Lumber*, in random sizes, is graded by the amount of usable material, of given dimensions, that can be cut out of the piece. The first and second grades, for instance, most often grouped under "FAS," yield between 83 and 92 percent of clear cuttings of 6-inch minimum width. In descending order the grades are *FAS*, *Selects*, *#1 Common*, *#2 Common*, *Sound Wormy* (allowing wormholes and limited sound knots), *#3A Common*, and *#3B Common*. Inspection is made on the poorer side of the piece.

Dimension Parts are meant to be used in the sizes provided for furniture or trim or structure. The grades are *Clear 2 Faces*, *Clear One Face*, *Paint*, *Core*, and *Sound*.

Finished Market Products include lath, siding, ties, planks, trim, molding, stair treads and risers, and flooring. Flooring is graded separately, in its own categories, by its own authorities. There is *1st Grade Light Northern Hard Maple*, *1st Grade Amber Northern Hard Maple*, *1st Grade Red* (which is beech or birch), and down to 2nd and 3rd grades, which allow more imperfections. Oak flooring is separately graded, and that in separate quarter sawn and plain sawn grades.

Softwood has its own two divisions, wood for *construction*, to be used in the thicknesses and widths delivered, and for *remanufacture*, to be

Lombard log hauler

Beyond the images of divinity, a great tree is a benchmark of life. We can point to the oldest living thing on this earth, a single bristlecone pine in the Inyo National Forest of California's White Mountains, over four thousand years old. It continues to survive in its arid environment by economy, by keeping alive only the thin strip of bark and sapwood and needles necessary to sustain life and aban-

reworked. Construction lumber is sorted into *stress-graded* (beams, decking, stringers, posts, timbers), *non-stress-graded* ("Yard Lumber," boards, lath, battens, planks, etc.), and *appearance* (trim, siding, flooring, paneling, ceiling, shelving). Stress-graded lumber is sorted as *construction, standard, standard & better* (std & btr), *utility,* and *stud.* In non-stress-graded lumber, imperfections such as knotholes, pitch deposits, and shakes are allowed to be larger and more numerous and the grade is lowered. In appearance lumber, letter grades reflect the quality of one face.

Softwood for remanufacture is graded according to the yield of usable pieces, as in hardwood factory lumber, and for specific uses (like ladder stock, molding, pole, tank and pencil stock). Factory grades descend from *factory select, select shop, #1 shop, #2* and *#3 shop. Industrial Clears* are for cabinets and doors. In redwood grading "All Heart" indicates good decay resistance.

Grading considers the species and the quality of lumber, and also its condition. The way lumber is surfaced, whether green or dry (less than 19 percent moisture), and how many sides have been surfaced are indicated by letter and number codes. "S1S2E" means "Surfaced 1 side, 2 edges." Other abbreviations listed here may help you decipher the cryptic yard inventory lists of a wood supplier.

Buying plywood is much simpler than buying lumber. Plywood is a standardized, regulated product, and a sheet carries on its face or edge a code that gives the user considerable information. Most softwood plywood is made

and graded in the United States, *Appearance* and *Engineered* grades.

Most sheets indicate the quality of face and back veneers, a *Span Rating* that suggests the minimum space between framing, the type of glue (*Exterior,* totally waterproof, and *Interior,* water resistant), often the product name and the code number of the mill that produced the sheet. The quality grades of outside veneers are: *N,* a smooth surface with a natural surface all heart or all sap that allows six neat repairs on a sheet; *A* is smooth, paintable, with up to eighteen neat repairs; *B* has a solid surface but allows minor splits, tight knots up to 1 inch, and circular plugs; *C Plugged* allows ⅛-inch splits, knotholes or borer holes to ½ inch, some broken grain and synthetic repairs; *C* has tight knots to 1½ inches, some knotholes to 1½ inches, synthetic repair, stitching, and discoloration or sanding defects that will not affect strength; *D* is utility stuff that has numerous knots, knotholes, splits, stitching, and is only for interior use.

Foreign-made hardwood plywood is not graded by U.S. standards. Most hardwood ply is for interior decorative use, but it is also structurally sound. Some softwood ply is available with a high-density or medium-density overlay that increases its strength, abrasion resistance, and gives it a smooth surface ready to paint. The American Plywood Association offers a helpful booklet explaining each plywood grade in detail, and additional books that may be available through your lumber dealer about plywood working, finishing, and plywood projects.

doning the rest of the wind-twisted, wind-scoured trunk, which stubbornly resists decay and remains to support the living part. The largest living thing on earth is the General Grant sequoia, a king among the barons of its brother sequoias at 279 feet high and 36½ feet in diameter at the base. It survives in its lush home by the enormous scale of its life and its three acres of roots. There are thicker trees:

Sawmills and Seasoning

by John Swain Carter

Crawling around sawmills or anyplace that has lots of machinery run on rubber bands, spit, and cussing provides an education that can't be found in any school. We often think of the craftsperson as being one who works in minute quantities, stingily searching for the ideal highboy, boat, or pot shape. This is fallacy. Craftsmanship includes other merits like efficiency, an understanding of tool capability, and an eye toward judicious use of material. A good sawyer combines all these traits and more.

The sawyer's tools can be as uncomplicated as a two-man saw or as complex as the giant automated western mills, where saws are run by computer and manhandling is nonexistent. Most commercial mills fall somewhere between these two. A large majority are small, often family run affairs catering to a variety of needs for local industries. These mills usually work on a contract basis and many (they are not enthusiastic supporters of advertising or Yellow Pages) can be sought out by questioning and keeping your eyes open while on back roads. They will invariably sell the fruits of their labor to anybody willing to pay cash and carry.

The quality of commercially available lum-

ber continues to decline and for most special applications, like boat building and cabinet-making, it simply isn't suitable. The alternatives are varied, each with an increasing degree of involvement in wood production. One of a craftsman's basics is a knowledge of materials and material processing. Involvement with the milling and harvesting process will aid a greater understanding of the material.

First the woodworker can go to a specialty lumber company or warehouse that sells various kiln-dried woods and purchase his stock from those available. Usually these firms buy lumber in lots through the wholesale market, kiln-dry it, and distribute it to specialty woodworking operations. Many leave the drying process to other companies and act simply as distribution centers. This approach to buying has advantages. It allows production to begin immediately with the excellent stock normally on hand. The major drawback is cost. High-grade kiln-dried lumber doesn't seem to be sold on the gold standard these days, rather it sets that standard. There are alternatives but they require planning — a word we often lose track of in a world revolving around convenience.

The prospective woodworker can also seek

BOARD FEET

SQUARE FOOTAGE OF PIECE (length' × width' = square footage)

THICKNESS	0.5	1	1.5	2	2.5	3	3.5	4	4.5	5	5.5	6	6.5	7	7.5	8	8.5	9	9.5	10	10.5	11	11.5	12	12.5	13	13.5	14	14.5	15	15.5	16	16.5	17	17.5	18	18.5	19	19.5	20
3/8"		.38	.56	.75	.94	1.13	1.3	1.5	1.7	1.9	1.9	2.3	2.4	2.6	2.8	3	3.2	3.4	3.6	3.75	3.9	4.1	4.3	4.5	4.7	4.9	5.1	5.3	5.4	5.6	5.8	6	6.2	6.4	6.6	6.8	6.9	7.1	7.3	7.5
1/2"	.25	.5	.75	1	1.25	1.5	1.75	2	2.25	2.5	2.75	3	3.25	3.5	3.75	4	4.25	4.5	4.75	5	5.25	5.5	5.75	6	6.25	6.5	6.75	7	7.25	7.5	7.75	8	8.25	8.5	8.75	9	9.25	9.5	9.75	10
5/8"	.3	.6	.9	1.25	1.6	1.9	2.2	2.5	2.8	3.1	3.4	3.75	4.1	4.4	4.7	5	5.3	5.6	5.9	6.25	6.6	6.9	7.2	7.5	7.8	8.1	8.4	8.75	9.1	9.4	9.7	10	10.3	10.6	10.9	11.25	11.6	11.9	12.2	12.5
3/4"	.375	.75	1.125	1.5	1.9	2.25	2.6	3	3.4	3.75	4.1	4.5	4.9	5.25	5.6	6	6.4	6.75	7.1	7.5	7.9	8.25	8.6	9	9.4	9.75	10.1	10.5	10.9	11.25	11.6	12	12.4	12.75	13.1	13.5	13.9	14.25	14.6	15
7/8"	.44	.88	1.3	1.75	2.2	2.6	3.1	3.5	3.9	4.4	4.8	5.25	5.7	6.1	6.6	7	7.4	7.9	8.3	8.75	9.2	9.6	10.1	10.5	10.9	11.4	11.8	12.25	12.7	13.1	13.6	14	14.4	14.9	15.3	15.75	16.2	16.6	17.1	17.5
1"																																								
1⅛"	.6	1.125	1.7	2.25	2.8	3.4	3.9	4.5	5.1	5.6	6.2	6.75	7.3	7.9	8.4	9	9.6	10.1	10.7	11.25	11.8	12.4	12.9	13.5	14.1	14.6	15.2	15.75	16.3	16.9	17.4	18	18.6	19.1	19.7	20.25	20.8	21.4	21.9	22.5
1¼"	.6	1.25	1.9	2.5	3.1	3.75	4.4	5	5.6	6.25	6.9	7.5	8.1	8.75	9.4	10	10.6	11.25	11.9	12.5	13.1	13.75	14.4	15	15.6	16.25	16.9	17.5	18.1	18.75	19.4	20	20.6	21.25	21.9	22.5	23.1	23.75	24.4	25
1⅜"	.7	1.38	2.1	2.75	3.4	4.1	4.8	5.5	6.2	6.9	7.6	8.25	8.9	9.6	10.3	11.0	11.7	12.4	13.1	13.75	14.4	15.1	15.9	16.5	17.2	17.9	18.6	19.25	19.9	20.6	21.3	22	22.7	23.4	24.1	24.75	25.4	26.1	26.8	27.5
1½"	.75	1.5	2.25	3.0	3.75	4.5	5.25	6.0	6.75	7.5	8.25	9.0	9.75	10.5	11.25	12.0	12.75	13.5	14.25	15.0	15.75	16.5	17.25	18.0	18.75	19.5	20.25	21.0	21.75	22.5	23.25	24.0	24.75	25.5	26.25	27.0	27.75	28.5	29.25	30.0
1⅝"	.81	1.63	2.4	3.25	4.1	4.9	5.7	6.5	7.3	8.1	8.9	9.75	10.6	11.4	12.2	13.0	13.8	14.6	15.4	16.25	17.1	17.9	18.7	19.5	20.3	21.1	21.9	22.75	23.6	24.4	25.2	26.0	26.8	27.6	28.4	29.25	30.1	30.9	31.7	32.5
1¾"	.88	1.75	2.6	3.5	4.4	5.25	6.1	7	7.9	8.75	9.6	10.5	11.4	12.25	13.1	14.0	14.9	15.75	16.6	17.5	18.4	19.25	20.1	21.0	21.9	22.75	23.6	24.5	25.4	26.25	27.1	28.0	28.9	29.75	30.6	31.5	32.4	33.25	34.1	35.0
1⅞"	.94	1.9	2.8	3.75	4.7	5.6	6.6	7.5	8.4	9.4	10.3	11.25	12.2	13.1	14.1	15.0	15.9	16.9	17.8	18.75	19.7	20.6	21.6	22.5	23.4	24.4	25.3	26.25	27.2	28.1	29.1	30.0	30.9	31.9	32.8	33.75	34.7	35.6	36.6	37.5
2"	1	2	3	4	5	6	7	8	9	10	11	12	13	14	15	16	17	18	19	20	21	22	23	24	25	26	27	28	29	30	31	32	33	34	35	36	37	38	39	40
3"	1.5	3	4.5	6	7.5	9	10.5	12	13.5	15	16.5	18	19.5	21	22.5	24	25.5	27	28.5	30	31.5	33	34.5	36	37.5	39	40.5	42	43.5	45	46.5	48	49.5	51	52.5	54	55.5	57	58.5	60
4"	2	4	6	8	10	12	14	16	18	20	22	24	26	28	30	32	34	36	38	40	42	44	46	48	50	52	54	56	58	60	62	64	66	68	70	72	74	76	78	80

Enter the table above with the square footage of the lumber's face and find its board-footage across from its thickness in inches

the Montezuma cypress *el Ahuehuete de Santa Maria del Tule* in Oaxaca, Mexico, is 40 feet in diameter three and half feet above ground level; it is only 135 feet tall, but it casts 8,600 square feet of shade at noon. A single banyan tree can send out its shoots and make a self-contained, wholly connected forest covering seven acres.

out sawmills in his locality. Some mills have filing systems, secretaries, and letterheads. At other sawmills the sawyer will do his figuring on the side of the mill and the final price results from prolonged dickering between the sawyer and customer. Some mills have lumber piles a cubist would envy and don't allow culling. Others have boards and logs everywhere in a symphony of confused order.

Most sawyers I've known come from a family tradition of milling and know the current value of a stick of pine as well as they do the value of a standing tree and the number of board-feet of timber it contains. If you're going to buy wood from these men an idea of the methods they employ is useful. Sawyers run their mills in two ways: custom sawing logs to the owner's specifications for a fee, and speculating on trees either standing or in the log. Payment on custom saw work can often take the form of a percentage of the lumber derived from milling.

Lumber is usually sawn two ways: plain and quarter sawing. Plain-sawn lumber is exactly what the term implies: the log is put on the saw carriage, squared, and then different-width boards are simply cut as the carriage makes progressive passes by the saw blade. The resulting boards are useful for most types of work, provided they are dried properly. Because it is cut with the annual growth rings running tangentially, plain-sawn lumber shrinks and swells more, due to the cell structure of wood. Heavy woods, such as oak, shrink more across the grain than lighter species, like cedar, and all plain-sawn boards

have a tendency to cup in the direction of their annual growth rings with changes in humidity. These effects may be undesirable in some types of work, such as fine cabinetry.

Quarter sawing produces a more stable board but also creates more waste and difficulty in sawing. Quarter-sawn, rift, or edge-grained lumber is sawn along approximately radial lines. The annual rings are usually no closer than a 45-degree angle to the face of the plank. Some change, often unpredictable, will always occur because of the organic nature of wood. Quarter- or radially sawn lumber, by virtue of its production and waste, is inherently more expensive than plain-sawn lumber. Both types mentioned above have to be kiln-dried or air-seasoned before they can be utilized for most projects.

Such monarchs speak to us of the mysteries now as if we were still primitives, still tree dwellers, but even local specimens have awesome proportions. Ages of one hundred and three hundred years are not uncommon. Heights of one hundred feet are not surprising. Many back pastures are dominated by oaks that spread a galaxy of leaves three times broader than their heights.

Some mills have their own kilns, others air-dry their lumber, and some sell primarily green stock. It is here that judicious planning on the part of the craftsman pays off. Green lumber from the sawyer can often be bought at a savings of up to 75 percent over custom lumber house prices. This green wood can be stacked and air-dried for use at a later date.

Air-drying wood is a matter of following a few rules, some patience, and much common sense. Wood is hygroscopic, which means it takes on or gives off moisture until it reaches equilibrium with the surrounding medium. The bulk of wood's moisture is contained in the cell cavities and is known as free water. This water is the first driven off in the drying process. The

fiber saturation point is reached when all of the free water is removed and the wood approaches a moisture content of approximately 30 percent. After this point the wood shrinks until it approximates equilibrium with the surrounding atmosphere. One rule of thumb in air-seasoning is to allow a year of seasoning to each inch of thickness. Many variables affect this equation: local weather conditions, stacking methods, air circulation, and cover prolong or shorten air-seasoning times. Balance is of primary importance in the air-drying process of timber. Seasoning should be slow and even. The essential features for stack drying are shown in the accompanying illustration. It is important to place hardwoods at the bottom

PEAVEY

CANT HOOK

SKIDDING TONGS

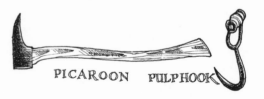

PICAROON   PULPHOOK

BUSH HOOK
LUMBER CARRIER

It would be enough if they gave only their shade and moisture and the color of growing, but trees give to us beyond the measure of any other thing with which we share the earth. I sit writing of them in a wooden house, in a wooden chair, warming my oak-tanned boots before a maple wood fire, drinking a kola nut beverage, munching walnuts and dates.

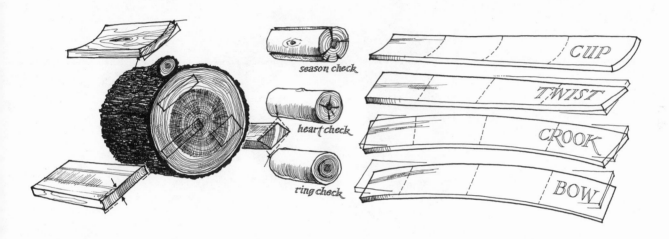

Effects of shrinking and warping

of the stack; stickers should be placed at approximately three feet; "chimneys," or unobstructed vertical passages, should be spaced throughout the stack for better air circulation; and the ends of the boards should be painted with a thick paint to help promote equal drying. A cover is recommended and can be formed in a pinch by pitching and overlapping the top layer of boards in the stack.

Once the lumber has dried to atmospheric level the boards should be stacked in the shop for some time before use. Warping and splitting could still occur if this procedure isn't followed before working the wood, because a high percentage of moisture is still present and should be allowed to equalize. There are various electronic gadgets that measure moisture content, but usually your judgment can determine if a board is ready for use. Feel the weight and compare, cut across the grain and check apparent moisture content. Most cabinetmakers consider 12 percent to be the ideal, but this is rarely met except under extraordinary conditions. Also, this percentage changes with application. It is possible to approximate the ideal, save money, learn more about your materials, and have excellent local stock to work with if these methods are followed and you are willing to make planning an integral part of your woodworking.

Spices, scents, and flavorings entertain and delight us from the forests of the world. The perfumes of sandalwood, camphorwood and eucalyptus are distant, complex, and rich. Sharp and fresh, a sprig of balsam warmed under a sleeping bag fills a poor tent with enviable fragance. No woodworker ever forgets the first time he sawed into rosewood, but

Further information on sawmilling:

Historical:
H. L. Edlin, *Woodland Crafts in Britain: An Account of the Traditional Uses of Trees and Timbers in the British Countryside* (North Pomfret, Vt.: David & Charles, 1973).

Alex W. Bealer, *Old Ways of Working Wood* (New York: Barre, 1972).

Useful:
Alexander J. Panshin et al., *Forest Products* (New York: McGraw-Hill, 1962) (2nd ed.).
A book on forest products technology. Each selection considers a part of the wood industry, such as: cooperage, cabinetmaking, and pulp products.

A. E. Wackerman et al., *Harvesting Timber Crops* (New York: McGraw-Hill, 1966) (2nd ed.).
An undergraduate textbook that covers basic harvesting techniques and the economics involved.

J. R. Dilworth, *Log Scaling and Timber Cruising* (Corvallis, Ore.: Oregon State University Book Stores, 1976).
This book is a must if you plan to have your own trees harvested. It explains the use of the log scale and cruising stick, both of which measure board feet of a tree either standing or cut.

Alexander J. Panshin and C. DeZeeuw, *Textbook of Wood Technology* (New York: McGraw-Hill, 1970) (3rd ed.).

An extensive text on the properties, structure, and formation of wood. It is more a textbook than a practical book for the woodworker but it nonetheless offers a compendium of basics.

Another source of information is the government. The Forest Products Laboratory publishes a number of books, booklets, lab reports, and fact sheets on all phases of wood technology. The Government Printing Office also publishes information on this topic. Ask for lists of works on a particular subject.

U.S. Department of Agriculture
Forest Products Laboratory
P.O. Box 5130
Madison, Wisconsin 53705

Superintendent of Documents
U.S. Government Printing Office
Washington, D.C. 20560

I first met John Swain Carter when he was mate and shipwright aboard the Clearwater. Now he is marine curator at the Peabody Museum. Because of his talent, experience, and scholarly knowledge, his position is proper and fitting. Most of us who have played darts with him, made wood chips, disposed of Pepsi and Barbados rum, and sailed and sung with him snicker at the thought of John behind a desk.

---

the more pedestrian timbers, too, have their beauties. The sweating shipwright is surrounded by a sweet musk of cut cedar and pungent Stockholm tar, and the plainest wood-butcher carries a sachet of hemlock chips and pine dust and fir shavings in his pockets and cuffs that surpasses the bergamot and amber-gris of Paris.

WOOD SUPPLIERS

Some of the sources of cabinet-grade lumber and rare woods.

American Woodcrafters
P.O. Box 919
Piqua, Ohio   45356

Amherst Woodworking
Box 464 Sunderland Rd.
North Amherst, Mass.   01059

Maurice L. Condon Co., Inc.
248 Ferris Ave.
White Plains, N.Y.   10603

Constantine
2050 Eastchester Rd.
Bronx, N.Y.   10461

Craftsman Wood Service Co.
2729 S. Mary St.
Chicago, Ill.   60608

Craftwoods, O'Shea Lumber Co.
York Rd. and Beaver Run Lane
Cockeysville, Md.   21030

The Dean Company
Veneers for cold molding, Olympic
   Manufacturing Division
P.O. Box 426
Gresham, Ore.   97030

Educational Lumber Co., Inc.
P.O. Box 5373
Ashville, N.C.   28803

John Harra Wood and Supply Co.
39 W. 19th St.
New York, N.Y.   10011

Homecraft Veneer
Dept. F, 901 West Way
Latrobe, Pa.   15650

J. H. Monteath Co.
2500 Park Ave.
Bronx, N.Y.   10451

Penberthy Lumber Co.
5800 S. Boyle Ave.
Los Angeles, Calif.   90058

Real Woods, Brookside Veneers, Ltd.
107 Trumbull St., Bldg R-8
Elizabeth, N.J.   07206

Rex Lumber Co.
P.O. Box 245
Cambridge, Mass.   02138

Robert J. Stalker
89 Pearl St., P.O. Box 307
Braintree, Mass.   02184

Sterling Pond Hardwoods
Route 100
Stowe, Vt.   05672

Violette Plywood Corp.
Northfield Rd. off Route 13
Lunenburg, Mass.   01462

Whistles in the Woods
Rte. 1, Box 265 A
Rossville, Ga.   30741

Wood Mosaic
P.O. Box 21159
Louisville, Ky.   40221

Wood Shed
1807 Elmwood Ave.
Buffalo, N.Y.   14207

"TREES WITH BUSHY TOPS TO . . ."

How much do you really want to know about trees and their wood? You can contemplate trees as part of the rural landscape or as a microscopic study or somewhere between the two extremes. In "Judging Distances" Henry Reed speaks for the army: ". . . Again, you know/There are three kinds of tree, three only, the fir and the poplar,/And those which have bushy tops to. . . ." There are books to satisfy the most general or the most specific wood appetite.

Identifying trees as they stand may be easy for the army but I have found the distinctions among them more difficult to recognize. The apples, alone, have 6,500 varieties, and 1,200 trees share the general name "palm." For rambles in the woods the Nature Study Guild's *Master Tree Finder* is compact and helpful, using leaves as the identification guide. They also have a *Winter Treefinder* using twigs, bark, etc. *The Golden Nature Guide to Trees* is also a fair companion for a walk.

Knowing Your Trees is a formidable book, presenting information, diagrams, and photographs that detail the trees of North America, their range, natural history, characteristics, and commercial uses. It is recommended, along with the United States Department of Agriculture's *Wood Handbook: Wood as an Engineering Material*, a prime reference for almost every commercially available wood, with numbers and specific information on strength, weight, and performance, as well as sections on building, fastening, fire treating, preserving, painting, and more. A bargain from the Government Printing Office, it's an essential part of your reference library.

The Constantine family have been involved in wood selling since 1812. Their long interest continues in the Constantine Lumber Company, dealers in imported and domestic hardwoods, and in *Know Your Woods* by Albert Constantine, Jr. It is a book that can fascinate and inform: I especially note the chapters on wood in medicine, on little-known application of wood, and the surprisingly interesting chapter on woods of the Bible. Even the wood quizzes are fun. I am not certain of how much use the descriptions of wood (listed alphabetically for 200 pages) can be to a craftsman confronting a slab of, say, Hinoki, but they may well encourage experimentation. Perhaps this is more a book for the library than the workshop.

The International Book of Wood is for the living room. It is the best mounted, most handsome, most comprehensive appreciation of wood's nature I have seen. The photographs in this work are first rate, but the illustrations are opulent, exquisitely detailed, colorful, and exciting. Highest recommendation.

Of two books that look at wood on a technical plane, *Inside Wood: Masterpiece of Nature* is the more interesting, beginning with visible growth rings and observable phenomena and reducing its focus to the microscopic workings of the cell wall and the chemical bonds in the atomic structure. There is much value in the chapters on use, decay, stress failure, and shrinkage. *Wood Structure and Identification* is a laboratory book, and though its election microphotography is quite beautiful it is of little practical value to the woodworker.

May Theilgaard Watts $1.00
Master Tree Finder   58 pp.
Berkeley, Calif.: Nature Study Guild, 1963   Illustrated

G. H. Collingwood and Warren D. Brush $7.90
Knowing Your Trees   374 pp.
Washington, D.C.: American Forestry   Illustrated
Association, 1937, 1965, 1974, 1975

Forest Products Laboratory $7.85
Wood Handbook: Wood as an Engineering
Material   Illustrated
Washington, D.C.: U.S. Dept. of
Agriculture, Forestry Service

Albert Constantine, Jr. $12.95
Know Your Woods   360 pp.
New York: Charles Scribner's Sons, 1954,   Illustrated
revised 1975

Martyn Bramwell, editor $29.95
The International Book of Wood   276 pp.
New York: Simon and Schuster,   Color photographs
1976

William M. Harlow $6.50
Inside Wood: Masterpiece of Nature   120 pp.
Washington, D.C.: American Forestry   Illustrated
Association, 1970

H. A. Core, W. A. Côté, and A. C. Day $12.95
Wood Structure and Identification   168 pp.
Syracuse, N.Y.: Syracuse University Press,   Illustrated
1976

Donald McKay $19.95
The Lumberjacks   319 pp.
New York: McGraw-Hill/Ryerson, 1979   Illustrated

A hearty book, *The Lumberjacks* is as pungent and satisfying as some of the eight-course loggers' breakfasts it
describes. There are people and tales and places in these
pages that have the ring and bite of life in the open air.
Good stories out of the deep woods seem — because of
The Lumberjacks — to be a literature smaller but as rich
as the literature of the sea.

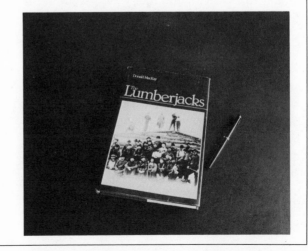

Every good kitchen is supplied with leaves of bay, with fruit of nutmegs, the bark of the cinnamon tree, allspice berries, and the dried flowers of the clove tree. The cacao yields cocoa and chocolate. The coffee berry, the tea leaf, and the kola nut are some part of cuisine. The sugar maple sweetens.

Sugarbush

by Perry Knowlton

Snow had fallen the previous evening. Not much. Three or four inches on top of the slowly melting base that, by the first week of March, had thinned to near nothing in the fields and on the roads where the sun could get at it, but was still substantial in the woods. "Sap snow," was what the farmers called it — this early spring snow which fell in the night when the temperature was in the high twenties — still below freezing — and was followed, as it would be today, by clear, sunny weather with temperatures in the forties and fifties.

Conditions were ideal, and we knew the sap would run heavily and that we'd be boiling in the afternoon. By noontime the pipelines above the sap house had half filled the big holding tank, which, at the turn of a valve, would fill the evaporator inside the house itself. There was a fine stand of hard maple on the steep slope above the sap house, and we had rigged our pipelines three weeks earlier. Each tree had two or three taps, and all of them, hundreds of taps, were connected by small plastic pipe to larger pipes and finally to a couple of one-inch plastic lines that snaked down the hill and fed into the holding tank. That tank held five hundred gallons, enough

maple sap to produce about twelve gallons of maple syrup.

And we had hung about seven hundred buckets in the good sugarbush on other areas of the farm and on the big maples that lined the dirt road skirting our boundaries. All of that had been done about two weeks earlier so we'd be sure not to miss any of the run once it started.

It was time to harness the team and collect the sap in those buckets before they filled and ran over. We'd collect twice today, given a little luck and a warm sun, which meant boiling well into the night. It would be long after dinner before we'd close down and let the fire go out under the evaporator.

Dick Liddle, the son of a local farmer, was now general hand and caretaker on my farm. He and Michael Doyle, a youngster from the Lower West Side of New York City who had become a close though unofficial member of the family, had rounded up some local kids to help with the collecting, and they had already brought Lady and Dolly to the barn. Dick took Dolly and Michael buckled on Lady's collar, hefted her harness off its peg and, laying the hames over the top of the collar, he heaved

Begin simply. After a juniper berry gin and quinine tonic (from the bark of the cinchona), dally with a bowl of walnuts and a glass of old sherry aged in oak. Move on to a cup of fruit salad — pears and oranges and peaches with a dusting of coconut and a splash of Cointreau. Cream of wild mushroom soup from the dim parts of the forest would be a mild introduction to small but perfect trout

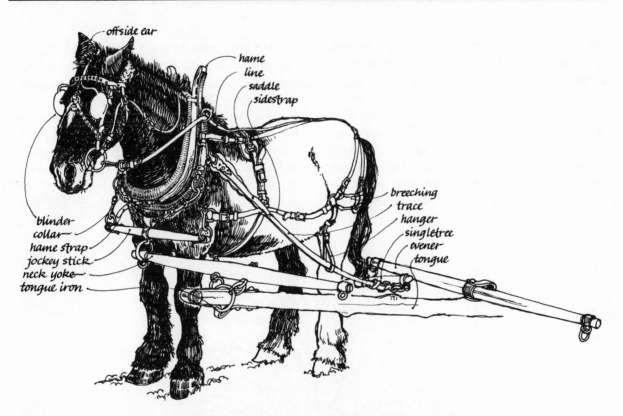

the harness up and over her back so the cinch dropped down on her off side where he could reach it under her belly, and he slid the breeching back over her rump. Going to her head, I bore down on the hame strap so the hames fit snugly to the collar, while Michael buckled the cinch to the near side of the belly band and pulled her tail out and over the breeching strap. In the meantime, Dick had done the same for Dolly and had already hooked the neck yoke to the bale straps on both horses' collars and was in the process of attaching the jockey sticks to the side straps.

Bridling both horses and setting up the lines were quickly accomplished, and with a quiet "Get up," I took the lines and started the team out of the barn.

The other kids had already lashed the collecting tank to the manure sled, a low-slung, heavily built sled about ten feet long which we found worked well for collecting. It was easier to maneuver than a long sled, and its center of gravity was lower, so with a full tank it was more stable in side hill conditions.

Dolly, the off horse, stepped neatly over the tongue, and on command the team backed so

we could hook the traces to the doubletrees, while one of the boys lifted the tongue to slip it through the ring on the neck yoke.

Fresh like this, after a day or two of loafing in the pasture, the team was a joy to work with. They moved out briskly in the snow, high stepping and eager, as we headed up the road to the long ridge with the stand of old maples, some of which were four and five feet in diameter and carried five buckets apiece. We

slowed to a walk as the boys jumped off and retrieved bucket after bucket and dumped their contents into the collecting tank. Occasionally we'd stop to let the kids catch up, and within an hour we'd finished the ridge, filled the two-hundred-gallon tank, and were headed back down the road, past the house and barn, toward the new sap house. The team pulled as if they weren't hooked to anything at all. A manure sled with over a ton of sap and five

people riding on it, and they clipped along as though they were pulling a round-back cutter full of feathers. And taken together they weighed a mere three thousand pounds. A pair of "farm chunks" is what they were, and good ones, too. The best team I'd ever had. So long as they had a breather now and then and a full scoop of grain for lunch, they'd do anything asked of them until dark, and after.

By the time we'd reached the sap house, the fire had been started under the evaporator and it was roaring away, stoked every few minutes with split firewood from the huge stack under the eaves. The evaporator had started to steam some, but it hadn't started to boil, so the house was still relatively clear. Before too long the air would be thick with steam. I backed the team so the drain pipe could reach the holding tank, fitted a screen over the end of the pipe to filter the sap, lowered the pipe, and drained the collecting tank; Dick Liddle stayed to supervise the boiling, and the rest of us climbed aboard the sled to continue the job of collecting from the trees not hooked up to the pipeline.

We headed for the sapbush at the bottom of the ridge on its south side where the sap ran best in conditions like these and filled the tank before we were halfway along the area we had tapped. We hauled it back to the sap house, strained it into the holding tank, and returned to collect the rest. So it went until we'd emptied all the buckets on the property itself, and all those we'd hung on the mile and a half of road leading to and bordering the farm. The way the sap was running, we'd be collecting

again by midafternoon, and we'd be boiling past midnight.

The big evaporator had been boiling all the while, and Dick had been checking to be sure the automatic feed system kept adding fresh sap as the liquid in the pan reduced. As the level in the evaporator dropped, float valves opened and fresh sap was supplied to the system from the holding tank. The evaporator itself was a maze of compartments and valves, which looked complex but was actually simple in concept. Fresh sap enters the first stage and travels gradually around a series of baffles that guide the liquid slowly over a fiercely hot fire below. At the end of the first stage, Dick Liddle (or whoever was boiling at the time) would occasionally allow the boiling sap in the first stage to flow into the second stage where the level had become lower through evaporation. There were three compartments in the second stage with simple gate-valves separating them, the same huge fire raging underneath. By opening and closing these gate-valves at the proper time, Dick was able to thicken the syrup as it passed from one compartment to the next. When the syrup reached the proper thickness, which we would know from the observation of a built-in thermometer and a hydrometer made specially for sap operations like this one, we would "syrup-off" from the last compartment. When canned at the proper time, a gallon of maple syrup will weigh eleven pounds, and it will neither spoil nor crystallize.

The sap house was full of steam. The hinged

smoked over hickory, for which a Spanish oak cork has kept an Auslese Goldtropfchen fruity and bright. Apple, pear, or peach cider for some. Next, I believe, some grouse from the upland woods, stuffed with wild rice and pine nuts, along with a smoked turkey and . . . what? A Pouilly-Fuissé, perhaps, a French vine grafted to American roots. A hearty main course: beef foddered on candletree for filet mi-

shutters in the cupola were raised high and windows and doors were open below, but it was still difficult to see across the evaporator to the other side of the sap house. Steam billowed up in great clouds, and the wondrous fragrance of boiling maple sap was everywhere. Once you've smelled a working, steaming sap house, you never forget it.

Dick checked again with the hydrometer. "That's about it. Let's syrup-off." A clean milk can was placed under the spigot, and a heavy filter made of felt and shaped like an upside-down ten-gallon hat was suspended from two dowels across the top of the milk can. A lighter and finer filter liner was fitted inside it, and the spigot was opened. When necessary we shut down the spigot, and changed filters, until we had almost drained the compartment.

gnon wrapped in hickory-smoked bacon and broiled over a charcoal fire. A hearty Burgundy with this, and a green salad with avocados and olives, dressed with apple cider vinegar and golden olive oil. Dessert is too difficult to decide; we must have a whole trolley of sweets: baked apple, tapioca from the cassava root, plum pudding garnished with holly leaves, berries of all kinds, cherry torten

---

31

It had reached about 240 degrees Fahrenheit when we syruped-off. We had set up a ten-gallon cylindrical tank on a framework over a space heater so that we could maintain that temperature for as long as we needed while we filled and capped the containers we were using, which ranged in size from six ounces to full gallons.

Once we'd started making syrup, everyone pitched in. Some washed filters, others helped pack the finished syrup, while Dick and I checked the evaporator, stoked the fire, and regulated the flow of sap from first entry to the last stage and final syrup. Fresh eggs from the chicken coop had been placed in a collander and hung in the boiling sap, so there were hard-boiled eggs and beer for those who were hungry and thirsty. "Jack wax" was a favorite of the younger kids. They'd ladle some of the thickening syrup on fresh snow, and the result was a sweet and chewy taffylike substance with the taste of maple sugar.

And before we knew it, it was time for collecting again. The horses had been fed and rested, and they were as eager as they had been in the morning. We left enough of a crew to keep the work going in the sap house, and we started over the same route we had covered earlier. The buckets weren't brim full this time, but most of them were over half full, so it was worth collecting again. To leave the sap overnight would have been taking a chance on its spoiling, though the temperature would probably be low enough to prevent that. In any case, it would be better to take advantage of the existing fire and to keep boiling until our

supply of sap ran out and to start out with empty buckets on the trees come morning and the new run of sap. Fresh sap is clear like good spring water and makes fine, light syrup. Spoiled sap is cloudy and sometimes yellowish, and although it's usable if not too badly spoiled, the syrup it makes is dark and not so tasty, and it doesn't bring as high a price in the marketplace.

It was after dark by the time we were back with the last load, and lanterns glowed through the steam in the sap house. You could smell the sugary steam half a mile downwind, and we could feel the temperature dropping fast now that the sun was down. It should be another good run tomorrow.

Once we had unloaded our last tankful, I sent three of the boys back to the barn with the horses, and I stayed to help with the boiling. Three of us were enough to operate the sap house now that the routine was established, so the others could unharness Lady and Dolly, feed them, and bed them down for the night. Then they would have dinner at the house and come down to relieve us so we could eat in our turn. The evaporator was full and boiling as it should, and the holding tank was just short of overflowing. It wasn't long before all of us had finished dinner and were back together working into the night. The night was getting colder, and the sky was full of stars, brilliant and sparkling from one horizon to the other.

The work went smoothly. The kids stoked the fire with the split wood from the seemingly endless pile; they skimmed the evaporator, syruped-off; and they packed the plastic and

in nut-flour pastry, honey-and-nut baclava, grapefruit with maple sugar, and, of course, black coffee steaming beside amber tea. The perfect end? Champagne, with almonds, pecans, and hickory nuts, and cherries covered in bittersweet, almost black, chocolate.

After that gorge you might need the bark of the cassia or the cascara, two mild laxatives in the pharmacopoeia of trees. Folk medicine

metal containers with the finished syrup. The gallons, half gallons, quarts, and pints were stacking up on the long shelf on the west side of the house. Once capped they were laid on their sides so the underside of the caps would self-sterilize in the hot syrup.

Working together, time went swiftly and before we knew it our holding tank ran dry, and with one last syruping-off, we opened the draft, front and back, on the firebox and let the fire burn itself out while the evaporator gradually cooled. While the fire died down we washed up the equipment, the hydrometer cup, the felts and the liners, the dippers, the skimmers, and the other odds and ends that had been in contact with the sweet syrup.

Before heading back to the house we restacked and counted the containers we'd filled. There were ten one-gallon cans, twenty half-gallon cans, eighty quarts, and eighty pints — a total of fifty gallons of maple syrup, all of it light amber in color. Pretty good for a small operation like ours. Fifty gallons of syrup represented two thousand gallons of sap.

And the dying fire and cooling evaporator again made us conscious of the falling temperature. It couldn't have been more than 25 degrees when we closed up and headed for the house. With the weather holding clear, the sun would warm things up quickly in the morning, and the sap would run again.

Perry Knowlton is a figment of his own imagination, who traded being a farmer of the 1880s (where he ploughed and planted and sugared with horses) for being a packet boat captain (he is the master of the 60-foot Hand motor-sailer Nor'easter). He has also flown airplanes, sailed ocean races, and been an editor for Scribner's. In the few days he spends off Nor'easter he is the president of Curtis-Brown, Ltd., a literary agency. He is, incidentally, my agent, my skipper, and my friend.

has always concerned itself with remedies from the forest and the stream lip — ash bark for fevers and agues, slippery elm for throat rasps, pine resin for coughs, and sassafras for the spring — but more serious remedies, powerful modern drugs, have come from the old herbalist's gathering of leaves and twigs. Studying Burmese folk medicine, pharmacopoeists developed a successful treatment for

Wood for Knife Handles

by Francine Martin

When a knife maker considers wood for a knife handle he wants a wood that is visually beautiful with distinct grain patterns fine enough to show up on small pieces. He also wants a wood that is hard and dense enough to stand up to heavy use and moisture. He wants a wood that is stable, shrinking and expanding as little as possible when taken from one climate to another. Obviously, a wood that develops small cracks while being worked or when changing environments will not be suitable for a knife handle. Finally, the knife maker wants a wood that is relatively easy to work, mills easily, and doesn't gum up belts too badly. Often, workability is not necessarily related to hardness.

The softest wood we use for knife handles is Indian rosewood. We use only the hardest woods, so when we talk about hardness, denseness, or the amount of resin, it is relative to other hardwoods in this category.

These (for the most part) exotic hardwoods tend to be very idiosyncratic and vary from tree to tree, and even within the same log. Grain patterns also vary greatly, depending on how the wood is milled. Exotic hardwoods are so inconsistent that we can only present

tendencies or generalizations that we have come to after working with these woods for many years.

Certainly you should wear some kind of mask when you sand these woods. Although many of them smell as sweet as your flower garden on a warm summer's day, the dust from most of them is poisonous to some degree.

We will discuss the individual woods in the order of their hardness, starting with the hardest wood in the world, lignum vitae, and working down to Indian rosewood.

Lignum vitae is so hard and resinous that it is used to make bearings for ships and ice-cream makers. It is almost totally resistant to corrosion of any kind. It has been found intact after 300 years in the submerged wrecks of Spanish galleons. It grows in Central and South America, and we use the heartwood only. This wood is very beautiful, visually and to the touch. It varies in color from a light golden honey color to very dark brown (rarely). Sometimes there is an iridescence that looks like crystallized honey. It very often oxidizes to a distinct green color. The grain is straight, tight, and nonporous. When polished it feels like a smooth stone

leprosy from the chalmoogra tree. Quinine from the cinchona treated malaria and played a part in opening the tropics. Today, quinidine from the same tree is used in heart conditions. The Indian rauwolfia contributes tranquilizers to modern psychiatric medicine. From the earliest studies and searches for drugs we have a range of poisons that are part of modern heal-

and cannot be dented with the fingernail. Lignum vitae is the ultimate in workability: it is waxy and appears to be self-lubricating, so it saws easily, is easy to shape, is fast cutting, and allows a long belt life. It is so resinous that it bubbles when heated. It has a pleasant cocoa smell when being worked. Unfortunately, this wonderful wood is quite unstable and tends to check; it also shrinks and expands readily in climatic changes. It is very brittle.

Partridge wood, too, is a tropical hardwood and fairly rare. It runs from a reddish medium brown to a very dark chocolate brown, with small, light-colored lacy grain patterns that look like partridge feathers. It oxidizes to almost black in use. We use the heartwood only of this relatively porous, open-grained wood. It doesn't gum up belts, but it burns fairly easily. This could be due to the fact that it's relatively nonresinous. It also has an unusual characteristic: it seems to glaze itself while being worked, which makes the belts feel as if they're being gummed up; but if you sand another kind of wood, the belt works just fine. Partridge wood is the most unstable wood that we have used.

African blackwood is about the same hardness as partridge wood; we use the heartwood only. It is black with translucent bronze swirls; there is a deep inner glow in the lighter areas. It is moderately porous and the most fragrant wood that we use — rosy and sweet. It resaws easily but gums up the belts with its tarlike

resin. Its color doesn't oxidize, but it is fairly unstable in climate changes.

Desert ironwood is one of the few nontropical woods that we use. It comes from the deserts of California, Arizona, and Nevada. The wood is quite rare, and it is illegal to harvest it in California. It ranges in color from dark brown (almost black) to a warm gold — sometimes in the same small piece. The grain pattern can be straight or tightly swirled and it sometimes has an inner translucence. There is no sapwood. Ironwood is not porous, but there is a distinct variation in grain hardness, so the finished handle is sometimes ridged. The wood is very difficult to resaw because it comes in gnarled and gritty chunks that are usually filled with sand and rocks. It is fairly easy to shape except for a slight tendency to burn. It gums up the belts just enough to aid in control. It is extremely resinous (though it doesn't appear to be) having a higher resin content than lignum vitae. This superb wood is *very* stable — almost rocklike; I have never seen it shrink or expand. Although it doesn't check while being worked, it is hard to find a good piece for your handle because of many existing checks. It has a distinct, effusive, and sweet chocolaty smell. The color does not oxidize.

Pau brasil, another South American hardwood, is vermilion colored, with an iridescence that changes from red to glistening orange depending on how you turn it toward the light. The grain pattern is straight and indistinct although there is a fine inner grain pat-

ing. The South American dart poison curare, for instance, is used as a muscle relaxant. The balm of Gilead tree soothes burns, the eucalyptus soothes bronchitis, the coca soothes pain, and the Jamaican dogwood soothes the spirit.

tern within the larger one which is almost geometric — regular, rectangular little soldiers all lined up in rows. It is easy to mill, but gums belts somewhat. It doesn't burn, nor is it excessively resinous. It is moderately open grained. The fragrance is somewhat reminiscent of eucalyptus, but sweeter and milder. It is very stable and the color does not oxidize; it keeps its brilliance almost indefinitely.

Osage orange is a most unusually colored domestic heartwood. It varies from lemon yellow to warm gold on the end grain and the cross grain is yellow. It is fairly straight grained, dense golden areas alternating with relatively soft and porous strips that are lighter in color. Some say that osage orange can oxidize quite readily to a light tan color. The color difference could be due to different origins of individual pieces or how the wood is finished or treated. It mills easily and gums the belts somewhat. It has a tendency to burn easily to a medium red color. I think that any red areas add to the beauty of the wood, but they do reflect imperfect control of the craft.

Kingwood, another South American hardwood, is close grained, with a very distinct grain pattern — basically straight with occasional swirls. It has mauve and purplish-black stripes, which oxidize slightly to the same basic colors with a brownish cast over them. It is sometimes iridescent, especially at the swirls. It is easy to saw, does not gum the belts or burn, nor is it excessively resinous. The smell is very heavy and sweet, becoming noxious

after a short time. It is clearly one of the more toxic woods we use. It is moderately unstable.

Purple heart. We use only the heartwood of this tight, straight-grained wood. It is reddish beige to mauve when first worked, oxidizing quite rapidly to lavender. The color and appearance of the grain patterns change according to the way the light falls on the wood, somewhat as with pau brasil but with more subtlety. It is easy to mill and to sand, except for its tendency to burn to a deep purple. (As with osage orange, I believe that the burned areas add to the beauty of the wood.) Purple heart is relatively nonresinous and does not gum up the belts. It is very light in weight relative to its hardness.

Coca bola comes from Mexico and South America; we generally use the Mexican heartwood. Most coca bola is brilliantly colored, varying from shades of purple through reds to oranges, yellowish brown, and black. There is also a wide variety of grain patterns, from the most intricate swirls to straight stripes. The hardness varies with color: browns and white are the softest; light orange and tangerine are harder; and dark orange, reds, and purples are the hardest. Coca bola resaws easily and doesn't gum the belts; it is relatively nonresinous. It is difficult to work because it tends to sand down more quickly than the blade steel. It has an oddly soft feel as it's being worked, even though it is a very hard wood. The smell is pleasant at first, but after working with the wood awhile the smell becomes noxious and

Wood's four great gifts are these: fuel, shelter, communication, and chemicals.

Wood is, in a general view, a simple substance: a fibrous cellulose material bound by lignin, hemicellulose, and water. But in closer view the family of woods is an organic complexity whose complete nature continues to baffle the chemist. Varying not only from one species to another and from one time to an-

Kitchen and utility knives by David Boye

David Boye $7.95
Step-by-Step Knifemaking — You Can Do   270 pp.
It!   Illustrated
Emmaus, Pa.: Rodale Press, 1977

For most of my waking hours I'm accompanied by one of David Boye's small utility knives, graced by Francine Martin's engraving and hung with a Gerber steel in a Pony Express Saddle Company sheath. Essentially, it is because I need a good blade for my rough frontier life — how could I face an orange rind, without it? Perhaps it is because Boye's strong forms and subtle grace notes are portable sculpture, useful adornment. Boye is one of the few great cutlerers who does not make knives with which to stick people: his skinners, utility knives, cooks' knives, choppers, and carvers are tools.

I question David Boye's business sense in writing such a good book about making knives. The commentary, photographs, and plain drawings have convinced me that I can make a knife as good as his knives. *Almost* as good.

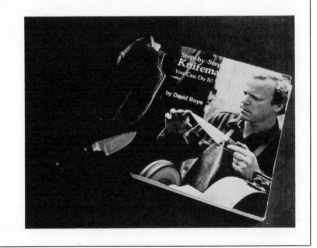

other, it also changes in chemical makeup from one part of the tree to another. The chemicals and chemical products from woodlots could fill this book and another, but if we consider only a few of the simplest substances we can see that their influence on civilization and technology is immense.

Iron fittings in green oak will be attacked and even destroyed by the high concentration

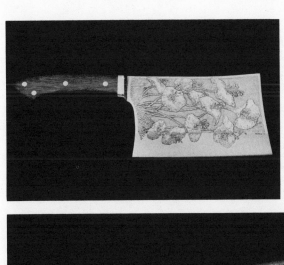
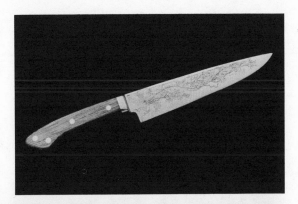
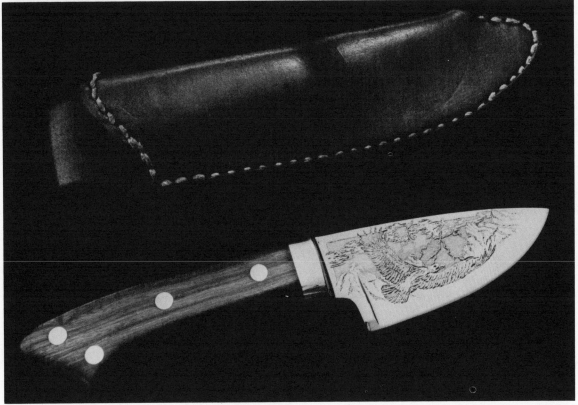

of tannic acid, the same substance used to tan leather, fix dye colors, and clarify spirits, and as a medicinal astringent. Bark has been collected for tannins for as long as men and women have worked leather. Resin and tars have been crucial materials for preserving, sealing, and waterproofing. Simple creosote is still used to keep timbers, posts, poles, and pilings sound in the damp earth, and against

Tool people often seem to be knife people. It's not surprising. A knife was the first cutting tool and it is astounding how much we still use them in a kitchen, on a boat, in the woods, on our own. The comfort of a knife doesn't necessarily lie in its weapon edge: it is more in its versatility, its use for many of the most basic needs, in some cases for its beauty.

Knives are very simple: that's one of the reasons they are so difficult to make well. The most difficult part is the blade, of course. Indian Ridge Traders sells blades and cutlery fittings at very good prices. Their catalogue ($1.00) is a short course in knife making, and they are pleasant folks with whom to deal. Their kitchen knives make up well, take and keep an edge, and their personal blades look good.

Indian Ridge Traders Company
Room 415, 306 S. Washington
Royal Oak, Mich. 48067

clearly poisonous; the dust is known to be quite toxic. It is very stable, but the color oxidizes to black when it becomes wet.

Brazilian rosewood, another South American heartwood, is black and brown with intricate pink and rich orange swirls. Some pieces look like coca bola. It is much more close grained than Indian rosewood. It mills easily and gums up the belts quickly, but it is easy to sand. When this wood is overheated it oozes a thick black resin. It most probably derived its name from its beautiful, sweet fragrance, which is especially pungent while the wood is being worked.

Tulipwood. A striking wood, it is moderately close grained with pink, red, and cream-colored irregular stripes. The red stripes are much harder than the cream. It is quite straight grained and I've never seen the color oxidize. A moderately resinous wood, tulipwood mills easily and does not gum belts. It works as if it were harder than coca bola, but it's not. (In other words, this wood does not cut faster than the steel.) It will burn to an orangish color if you use a worn belt. Like kingwood and coca bola, it has a sweet smell that turns heavy and noxious.

Indian rosewood has an overall purplish cast, and large, subtle grain patterns of black and purple, which tend to be straight, though sometimes slightly swirled. It is very open grained and porous, and very light for a hardwood. It is easy to resaw, but somewhat difficult to sand. It oozes a black resin when being worked, which badly gums up belts. This resin also forms a crust on the wood itself which is hard to penetrate while sanding. The wood has a very pleasant, light, sweet fragrance. It doesn't oxidize and is very stable.

The woods that I have discussed here are just a few of many beautiful and exotic woods. We have used others at various times but these have the right balance of characteristics for knife handles.

Francine Martin is a woman of great graphic talent, a shy manner, and a soft voice. She engraves the blades of David Boye's beautiful knives — owls, whales, damsels, foxes, seas, and mountains. It is a small shop producing sculptures for the hand, tools with edges, and hand-friendly handles, and to the best of my knowledge the firm does not make "fighting knives." Just as well.

the salt, salt sea nothing has rivaled Stockholm tar. Perhaps the most pivotal stuff that comes from the forest is charcoal. It is produced by burning wood without enough oxygen for complete combustion, sacrificing volatiles and trace substances, leaving a porous carbon structure. The porosity of carbon, with a surface area of several acres to the pound, makes it an ideal filtering agent for chemicals,

Burning Wood

by Jan Adkins

The woodworker can have mixed feelings about wood heat. On one hand he feels the continuum of the forest, the multiple generosity of the trees providing livelihood and lumber and food and heat and shelter and the paper to read about it. Add to that a woodworker's supply of kindling and it is natural that he should burn wood. On the other hand, his closeness to wood argues against burning good timber for heat. On the one hand a woodworker can be affected by the death of a beautiful tree, on the other hand coal and oil are such ugly commodities.

Hefting split logs into my stoves I often stop to admire a particularly straight, clean billet of maple or birch and reflect that I should drawknife it and chuck it on the lathe instead of burning it. Then I chuck it into the stove instead, but always with a twinge.

It can save you money, get you closer to your life and your roots, and offer an involvement with the elements you can't have with a thermostat and slave heat. You can even come to prefer the quality of heat it provides: a good stove gives a steady, even heat from a point source whose radiant glow changes on your face and hands as you cross the room and

makes chair cushions and wall surfaces warm to touch. The cold of winter is felt as much by your emotions as by your skin, and the stove makes a warm refuge where you can hunker down and be comforted in many ways. A hidden furnace, for all its BTUs, can't be the glowing heart of a family in the way a whispering stove on the hearth can.

The gritty details of wood heat, however, are not so lyric. It is not convenient, at least in late-twentieth-century terms when we define a refrigerator that doesn't drop ice cubes into your glass automatically as something less than the state of the art. Hell, in *early*-twentieth-century terms it's not that convenient. Cutting your own wood is a long, moderately hazardous, definitely strenuous chore. Your labors are not even complete if you have a woodcutter dump a cord and a half on your flower bed. You must stack it so it doesn't fall over on your car's hood. You must cover it, tightly enough so the sound of flapping vinyl doesn't drive you bananas on a windy night. You must carry it to the fire in quantities small enough to minimize bug and borer immigration and large enough to minimize mileage between hearth and woodpile. You must light it; no number of merit

Illustrations from *The Art and Ingenuity of the Woodstove* (Adkins)

air, drinking water, and sipping whiskey. Once ignited, charcoal burns with an extreme heat that made the smelting, working, and alloying of metals possible and set the human capacity for work and change on a higher plane. The alder and the willow have a special place in the history of change, for their charcoal is especially suited as a component of that

badges will help you here and there will be much gnashing of teeth of frosty mornings. You must tend it; this is a matter of many critical stares at midburn and a bit of poking or raking now and again (though a good firebuilder can tend to it every eight or ten hours). You must clean up after it (strenuously truthful, I admit that it is not a good way to heat an operating room); ash and dust from the firebox settle over the room, while the more noticeable detritus from logs and bark is confined to an area around the woodpile and a path to the door.

Does all of this sound discouraging? That is my intention. Not every household should burn wood. Our forests are renewable only to a limited extent, especially the hardwood forests that give us prime heating wood. Monoculture, the philosophy of the great wood enterprises, excludes slow-growing deciduous hardwoods in favor of crop trees, economically harvestable conifers. Defoliants are now assuring an "economic" future for many millions of acres. It is likely that woodlots all over the country can supply us with sufficient building hardwood — more land is forested today than in 1870 — but if whole states converted to wood heat the hardwood forests would be taxed beyond regeneration.

Combustion of wood is a simple process as chemically complex processes go, and its byproducts are fairly innocuous, but the large amount of particulate matter in a woodstove's smoke would, if multiplied by twenty or thirty thousand stoves in a meteorologically confined area, set up ugly and annoying and even hazardous situations. The best woodstove com-

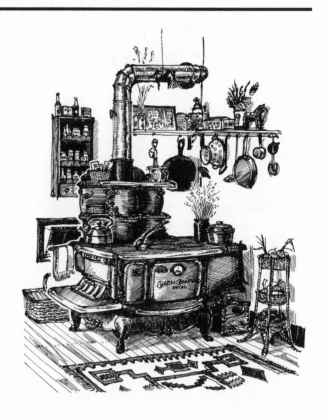

mercially available would not pass an EPA emissions test.

Before you decide to heat with wood, consider your abilities, location, resources, the state of your local environment and the national economy, your capacity for hard work, your inclination to sweep and dust, and your schedule at home. If you balance it in favor of wood heat, you are lucky: heating with wood is a rich experience; as for the drawbacks, I can only say that I am a villain of minimal character and I manage it.

There are several good books on wood heat that treat the whole subject thoroughly (among

Charcoal burners in Britain

them my own — many are discussed on pages 46–47). They will help you answer most of the questions wood heat poses and fill you in on the history and background of stoves. If you see to basics (the wood, the stove, and the installation), though, you will do well.

In the simplest view the wood you burn should be dense and dry. All wood yields very nearly the same amount of heat *per pound*. A cubic foot of light wood yields less heat than a cubic foot of heavy wood: this is the reason hardwoods are preferred for heating. For a given heating job, a house or a shop, a stoveload of black locust equals about two stoveloads of eastern hemlock; that's twice the hauling and twice the tending. At 12 percent moisture (fairly dry) black locust weighs 46.1 pounds per cubic foot, while hemlock weighs 25.9 pounds per cubic foot. Longleaf yellow pine (a softwood), at 37.7 pounds per cubic foot, gives more heat per stoveload than quaking aspen (a hardwood) at 24.5 pounds per cubic foot, but in the main hardwood is denser than softwood, and wood is not sold by weight but by volume, in *cords*.

The cord is a cunningly anthropometric measurement of stovewood, based on a woodcutter's capabilities. A cord is four feet wide, four feet high, and eight feet long: a log or limb 4 feet long is about the biggest piece a cutter wants to horse around with; four feet is about as high as a cutter can comfortably heft a four-foot log to stack it; and an eight-foot string of four-foot logs stacked four feet high is a morning's or an afternoon's work, with handtools. All other measurements get tricky: face

cord, stove cord, Wichita cord, swamp cord . . . disregard them and buy by the cord, 128 cubic feet in any proportion. Some discrepancy, as much as 8 percent, can be accounted for in kerfs if the wood has been cut into stove lengths, but it remains the only proper way to purchase firewood.

There can be major differences between two cords of wood of the same species. A cord of seasoned wood burns easier, hotter, and produces more heat overall than a cord of green wood. Green wood can weigh almost twice as much as seasoned wood, and that difference is water. This wet wood burns with less heat because energy is dissipated in vaporizing the moisture. "Dry" wood, less than 25 percent moisture, can make a quick, hot fire with a few sticks of kindling and a few sheets of newspaper, but a green wood fire is a frustrating affair that usually needs a healthy dry wood fire to start it and can smolder and die any time. Seasoning wood is a matter of making surface area available to air, of providing lively ventilation, and of protecting the wood from additional moisture. Splitting green wood exposes surface area and speeds seasoning. Stacking split wood loosely promotes free circulation of air all around. A covered, drafty shed is ideal for seasoning wood; it keeps off rain and snow and allows good ventilation. Covering a woodpile with plastic can be worse than leaving it exposed: the plastic shell can hold moisture rising from the ground. If you haven't a shed, cover the *top* of a stack with a strip of plastic weighted at the sides and block the wood up with concrete blocks, stone, or

nation-breaking, age-birthing admixture of carbon, sulfur and saltpeter: gunpowder.

Wood as a part of communication has brought knowledge to us from distant places and past times. Until the telephone and the computer net grows far more dense, information will be sent, stored, and inscribed as it has in China since the first century A.D. and in the Western World since the twelfth cen-

waste wood, up to six or eight inches if possible. Storing wood in your house or on your porch is destructive; the firewood's moisture and its load of termites and other wood eaters can damage your home's structure. In three to six months, under good conditions, the wood will season. You can mark a medium log and weigh it, keep it with the rest of your wood, under the same conditions. When the weight stabilizes for several weeks, the wood is seasoned.

The stove that burns your wood is as important as the species and state of your wood. A fireplace is a device for wasting heat. Wood is placed on dogs, lit, and burned, heating air rushing up the chimney and pulling one cubic foot of cold air into the room for every cubic foot of heated air it throws outside. The only heat it provides is radiant heat, a minor part of its expended energy. Enclose the fire in a metal box and take it out of its masonry cubby, out into the room, and you have a great advance. The stove allows air to flow around the conductive metal and promotes warm convective currents.

The next advance is to control the fire itself. The two parts of a fire are fuel and oxygen; load the fuel and control the flow of air (oxygen) and you control the burning rate. You can control the flow of air only in an airtight vessel, so the stove's joints and door must be airtight, allowing air to enter only through adjustable drafts. With such a stove a burn can be started quickly with a heavy flow, cut down when the room is up to temperature, and slowed even further to stretch the fuel overnight. An airtight

stove burns at an efficient rate, conserves fuel, provides a more even heat, allows adjustment of room temperature, and makes an overnight fire possible without banking or mixing in green logs. An airtight stove is the simplest and the most effective way to burn wood for heat.

In utilizing every available part of the wood's heat energy and in lengthening the practical time of a burning there are trade-offs and difficulties and even dangers. Hundreds of commercial stove designs contrive to burn longer and pass more heat from each pound of wood, a competition that obscures — often deliberately — the drawbacks involved. As a piece of wood burns it first expels some of its water, then gases (volatiles, burnable and unburnable), and finally it consumes the carbon (char-

tury — on paper. Today we have a system of stored facts and opinions that reaches to the smallest town library. Only recently have television and radio news seriously challenged the position of the newspaper as the broadcaster of current events, and newspapers place a major demand on the forest: one Sunday edition of *The New York Times* clears hundreds of acres of pulpwood. Even twigs have

a run

2'

4'

a unit

2"

18"

16"

a cord

8'

4'

4'

been used as messages. In the American Southwest notched tally sticks still record the number of sheep in flocks. Until 1826 the English Exchequer issued willow twigs as receipts for taxes; after a twig was notched to the amount of the tax paid, it was split lengthwise, and one side was given to the taxpayer, one side kept as a record. Presented as proof of payment, the split, bark, notches, and grain

coal) left. The smoke and vapors given off by burning wood are gases and particulate matter swept along with them and are mobile until their temperatures drop to a dew point, a temperature at which the gases condense into liquid. Water vapor in humid summer air reaches its dew point around a cold glass of iced tea and condenses on its surface as beads of water. When the temperature of smoke drops to its dew point it condenses on the inner surfaces of the flue as *creosote*, a complex compound, noxious, corrosive to mortar and steel, and gooey until it dries to a brown/gray/black scale that narrows the flue passage and presents a fire hazard. A narrowed passage is a bother, but a flue fire is dangerous. A long, vertical tube packed with fuel makes a pow-

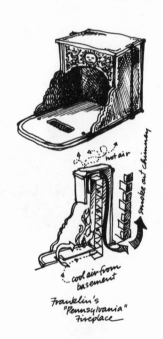

Franklin's "Pennsylvania" Fireplace

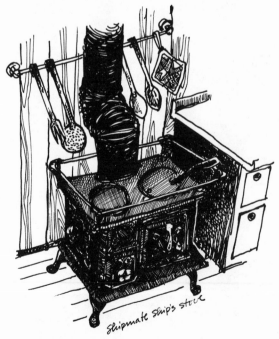

Shipmate Ship's stove

erful blowtorch: the fire flutters, violently shaking metal pipe heated to a cherry and then a white glow; in masonry the flame licks out of chinks and faults at concealed beams and rafters; a plume of sparks and burning cinders streams up into the sky and settles down on trees, roofs, and outbuildings. A chimney fire is an emergency.

Some homeowners sure of their flues (that they are lined with tile in good condition), the state of the roof, and the prevailing fire hazard for surrounding trees, intentionally start chimney fires regularly. This periodic burning away of creosote assures that any chimney fire begun accidentally will be small. The classic

had to match; a receipt impossible to counterfeit. When the system changed the twig files were burned. The fire ran out of control and burned down the Houses of Parliament.

Do not neglect the significance of the wooden ship in the history of communication, nor the minor note that the black ink in which I am writing comes from the oak tree: tannin and iron salts.

A simple Russian Fireplace

chimney calamity is the uncontrollable blaze from a year or many years of accumulated creosote, ignited by burning Christmas wrapping. Out of sight, out of mind.

Chimney fires can be avoided altogether (undoubtedly the safest plan) by removing deposits of creosote and minimizing creosote buildup. Once a year (more often if you burn over four or five cords a winter) the chimney should be cleaned by you or by a chimney sweep. You can use wire chimney brushes or one of the patent devices or a jumble of chains and scrap. It's a messy, awful job but it's as necessary as locking your door in the city. A professional sweep will give you an estimate

beforehand and, often, a free safety inspection.

Minimized buildup results from choosing, installing, and using your stove carefully. The firebox of your stove should operate at a very high heat to burn as completely as possible all the volatile gases. To ensure this extreme temperature many stoves have iron or firebrick linings that contain and concentrate heat and lessen the cooling conductivity of steel or cast iron in contact with the fire. These stoves pass the very hot flue gases on to baffles or other conductive devices outside the firebox, that give heat to the room without lessening the fire's operating temperature. How well these devices should work is questionable: the flue gases should be hot as they leave the chimney; if too much heat is removed from those gases, they will condense inside the flue and draft will be poor. In churches and schools of the last century it was common to see stoves with long flue pipes extending across rooms to place stoves at the center and to radiate heat from the flue itself. "Dumb stoves" were collections of thin-walled baffles on the upper floors of a house, connected to the flue of a stove on the ground floor; though they had no fire themselves (some had a small firebox, seldom used) they radiated heat by forcing the hot flue gases to move slowly through their baffles and release the heat to the rooms. Today, "stack robbers" — conductive/convective devices fitted into flue pipes above stoves, many with electric fans — similarly drain flue gas heat. In all these cases the result is increased creosote buildup, the increased need for chimney and

Not until the 1830s was geologic oil seen as anything but an annoying oddity, and it would not be used to heat homes until almost a century thereafter. Natural gas was a major heat source much later. Electricity as a home heater was not practical until our midcentury. Coal, the seam-laid remnant of ancient forests, and peat, its immature brother, have been important fuels for a century or three. For all the

WOODBURNING BOOKS

The first rumblings of the energy crisis are sending folks outside the city back to wood heat, and in the midst of the crowd are a babble of authors riding the wave. Since I'm among them I can speak of their harmless mixture of motives: concern, piety, and greed. Out of this stampede have come a few good books and several less useful. Unbiased as always, I recommend my own, but add that *The Woodburner's Encyclopedia* is an interesting companion. In general order of preference, then:

Jan Adkins	$12.95
The Art and Ingenuity of the Woodstove	136 pp.
New York: Everest House, 1978	Illustrated

A comprehensive and well-illustrated story of woodstoves and woodburning: burning theory, wood gathering and storing, history, chimneys, installation, and access. The only book on the subject that is, to my mind, readable and beautiful. More than a nuts-and-bolts guide (which it most certainly is), this is a book you will want to take with you even when you move to warmer climes in Arizona. I should sign another name to this review but I *like* the book.

Jay Shelton and Andrew B. Shapiro	$6.95
The Woodburner's Encyclopedia	155 pp.
Waitsfield, Vt.: Vermont Crossroads Press, 1976	Illustrated

Dr. Shelton is still at work generating the real numbers of woodburning: stack pressures and temperatures, BTUs per pound of fuel, insulation factors. This is an understandable and very interesting technical source. All wood burners turn into bores on the subject of BTUs, and the tables in this volume fuel the woodstove talk into a blaze of technocratic certainty. You'll need the book if only for self-defense against the cord-heads.

Mary Twitchell	$7.95
Wood Energy — A Practical Guide to Heating and Wood	170 pp.
Charlotte, Vt.: Garden Way Publishing, 1978	Illustrated

A pleasant and workable effort, understandable and broad, this book contains most of what you might want to know. Parts of this book ("Axes & Chainsaws," "Chimney & Stove Cleaning") are available as bulletins from Garden Way, at $1.00 each.

John Vivian	$4.95
Wood Heat	320 pp.
Emmaus, Pa.: Rodale Press, 1976	Illustrated

Presented in a personal, folksy style typical of Rodale Press, this book is understandable, if mannered, and its illustrations take a fairly broad approach. A nice section on woodstove cooking and some agreeable sidelights are found toward the end (soap making with ashes, wood-heated irons, etc.). I wouldn't construct the wood-and-stick-cattied chimney unless I was in the arson business, but it's the thought that counts. A friendly book.

Bob Ross and Carol Ross	$10.00
Woodburning Stoves	142 pp.
Woodstock, N.Y.: Overlook Press, 1976	Illustrated

This fairly extensive text was written from obviously considerable experience, but is marred by surprisingly bad drawings.

David Havens	
The Woodburner's Handbook: Rekindling an Old Romance	96 pp.
Brunswick, Me.: Harpswell Press, 1973	Illustrated

A small treatise and a minor effort. Hardly adequate.

time before that, for all the places where a family might need warming, and for the cooking of bean sprouts or caribou, there was one main fuel: wood.

Today many of us are alarmed at the limited amount of oil and gas left to us and are searching for alternatives to heat our homes and schools and business places, dry our crops, and power our cars. Some homes are being

Jane Cooper $5.95
Woodstove Cookery   196 pp.
Charlotte, Vt.: Garden Way Publishing,   Illustrated
1977

Giving substantial, congenial advice on buying, installing, firing, and using a woodstove as the heart of the kitchen, along with some smashing recipes, Ms. Cooper brings out the notable fact that woodstoves are especially effective in some styles of cooking and gives a delectation of stir-fry recipes that demand a rush to the kitchen. Among many, "Corny Waffles," "Lime-fried Chicken," and "Sesame Squash Delight" have me salivating like Pavlov's dog at a clock auction.

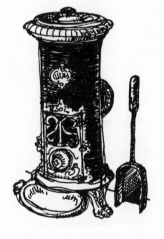

Robert Bobrowski $5.95
Rediscovering the Woodburning Cookstove   96 pp.
Old Greenwich, Conn.: Chatham Press,   Illustrated
1976

Those who hand-letter books should have a very good hand. This book suffers from Mr. Bobrowski's scrawl. His illustrations are almost primitive and yet have a directness and a charm that would have been enhanced by well-set type or a simpler, better calligraphy. As it stands, this book is an amateur attempt, but it does have some interesting woodstove details.

heated with wood again. This return to wood heat is significant in many ways: it marks an awareness of our energy dilemma and a willingness to be personally involved in it. Heating with wood points out in a way few other pursuits can the difficulties of making warm shelter and makes us mindful of our great-grandparents' struggle to make a good home; watching a woodpile dwindle through the win-

flue cleaning and for vigilance against fires, and the increased probability that during a chimney fire the violent shuddering of the pipes and devices will open a joint or pull loose a fastening and cause a house fire.

Stove makers and proud stove owners boast of fires kept for fifteen and even twenty-four hours. An airtight stove makes long burns possible, holding air flow to the minimum necessary for feeble combustion, but protracted burnings generate little heat and much creosote. The firebox temperature is low, the wood smolders more than burns, draft is weak and slow, the volatiles condense and collect in the cool stack. Eight or nine hours is a sensible burn, enough to present a stoveful of good coals in the morning or after work. Stretching it much beyond that is the kind of false economy you can brag about and regret.

Choose a stove to suit your location and room and style. If you live in the suburbs or the city consider one of the excellent coal stoves (Vigilant Coal model, Chappee, or the Petit Godin) as an alternative to woodstoves. Coal can be had from city suppliers, handled in bags and scuttles with less mess, and stored indoors without termite stowaways. Consider the size of the area you plan to heat and use a stove of an answering size. Most neophyte wood burners err on the side of overkill, installing a stove of prodigious capacity that turns their studio or living room into a sauna, backing them into running it at a slow, inefficient, creosote-producing burble. A small stove breathing hard to heat a room is preferable to a big stove loafing and smoking.

There are several schools of woodstove styling: the "Maximum Efficiency" or "Spacelab" school, the "Stark Utility" or "Euclid Earthmover" school, the "Decorator" and "Victorian Phantasmagoria" schools, and a few fine designs that stand on their own. You can choose a stove for efficiency, ease of operation, or style. Of these, the two most important factors are style and operating ease. A rationale for excluding efficiency claims is that most airtight stoves fall in a fairly narrow spread of efficiencies, say 10 percent, while the difference in efficiency between any two loadings of wood can be as much as 40 percent, even when the wood comes from the same pile. You might as well pick a stove that will be easy to load and

ter gives us a sense of natural urgency about weather and the seasons we may not have felt for generations; seeing wood as a growing tree, felling it, working it up into firewood, and loading it into the stove gives a sense of wholeness, a feeling of belonging to the cycle; heating with wood impresses us with our old reliance on the woodlot, our ancient, warm friendship with the trees.

light and be an attractive addition to the room as live with a great hunk of industrial iron for the sake of four or five percentage points of efficiency overruled by the inconstant nature of wood.

For many wood burners the charm of fire itself is a part of their pleasure and several airtight stoves oblige them by opening into fireplaces, for a story in front of the fire with the kids or a pan of chestnuts on the coals. If well made, these can be a good choice because they load through a broad opening.

A stove for a shop should match the shop's use. If the shop is used only sporadically and for a few hours at a time, the stove can be thin walled and simple, providing fast heat with a hot roar. If the shop is an all-day, everyday workplace, a heavy, large-capacity stove suits the need for long, even warmth and overnight burns.

Wood-fired cookstoves bring a new array of skills and disciplines into play for the art of cooking. The complicated, ornamented, antiquated cookstoves of old kitchens are being rediscovered in barns and cellars as treasures. Some gas-fired chefs grump that they are slow, balky, unpredictable, and uncontrollable. Some down-home stewcooks enthuse that they are gentle, even bakers and offer delicate differences in heat over their broad surfaces. The facts are probably somewhere between these views, but the signal truth is that the wood cookstove is a companion, warm and lovable in its cast iron excesses and old style amenities.

Installing a stove begins with deciding where it should stand. It should be near the center of the home, if possible, where its heat will be retained, and not on an outside wall where part of its heat will be drained off by the cold outdoors. It must have access to a flue or a flue must be built for it. If an old flue exists you must be certain that it is fit to continue in use. Old flues can be unlined, their mortar may have dried and fallen away, whole bricks may be missing, cracks and clear spaces may be open to beams and rafters outside the flue. These faults, combined with the corrosive effect of creosote, its flammability, and with the violent, ungovernable nature of chimney fires, are too dangerous to ignore. An old flue should be inspected, repaired, and perhaps lined with new mortar, with patent interlocking flue tile, or with screw-jointed metal pipe. If a new flue must be built, there are three choices; metal prefabricated insulated pipe, pipe, and masonry. There are considerable advantages to prefabricated flue pipe. It is simple to install, requiring only basic carpentry to pierce the roof and set up. It heats up rapidly to build a strong draft quickly. Its insulation, besides protecting adjacent flammables, contains the heat of flue gases and lessens creosote condensation. It *is* expensive, about $1.00 to $1.25 an inch, but not when compared to the cost of a foundation-to-height masonry chimney.

Local codes dictate the clearances around stoves and flues. Generally they are similar throughout the United States. Consult your building inspector, fire marshal (who may also help you inspect your flue), or a good stove emporium to find what local requirements are.

ANTHROPOMETRICS

We are told that within those minuscule computers, the DNA chains in our cells, there are more similarities than differences between us and earthworms. Though each human being is unique, our similarities are more striking than our differences. A designer of a piece of furniture, of a house or handrail or an outhouse . . . of anything to be used by more than one person . . . needs information on those similarities — sizes, heights, and clearances that accommodate most of us. Anthropometrics is a statistical discipline describing those dimensions and "standard" heights and placements almost arbitrarily decided but established by use and expected as a normal part of a room or setting — the heights of light switches, for instance, or counter tops or door knobs. Here are only a few, and there are thousands more. Other standards, like tennis-court dimensions, brick proportions, and filing-cabinet sizes can be parts of a design too. Some reference standards are:

Alfred Dreyfuss, *The Measure of Man* (Whitney Library of Design)
Architectural Graphic Standards (1926)
Time Saver Standards (New York: McGraw-Hill, 1966).

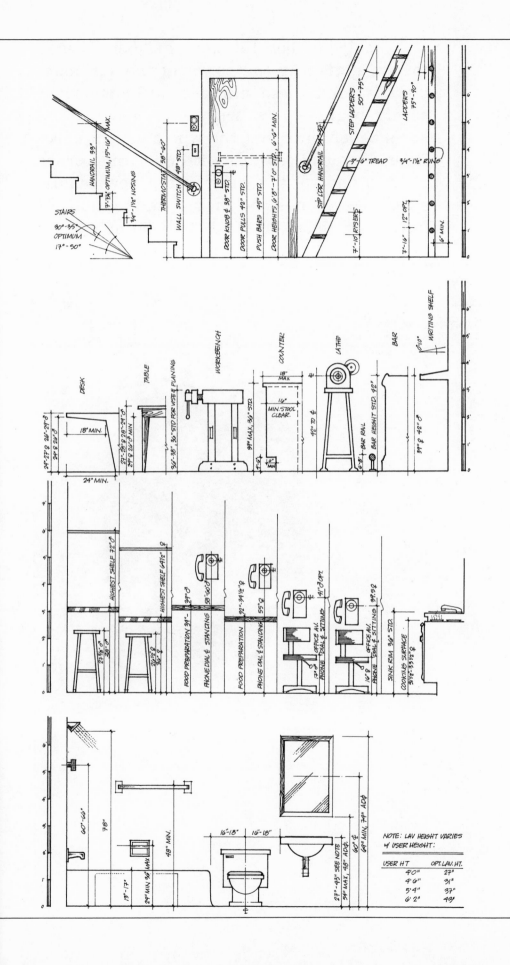

STAIRS
30°-35° OPTIMUM
17°-50°

HANDRAIL 33"
7"-7¼" OPTIMUM, 15"-16" MAX.
¾"-1½" NOSING

THERMOSTAT 58"-60"
WALL SWITCH 48" STD.
DOOR KNOB ₵ 38" STD.
DOOR PULLS 42" STD.
PUSH BARS 45" STD.
DOOR HEIGHTS 6'8"-7'0" STD. 6'-6" MIN.
STP. LDR. HANDRAIL 34"-38"

STEP LADDERS 50°-75°
LADDERS 75°-90°
3"-6" TREAD
¾"-1½" RUNG
7"-16" RISERS
7"-16" OPT.
6" MIN.
12" OPT.

51

DESK
TABLE
WORKBENCH
COUNTER
LATHE
BAR
WRITING SHELF

24"-27" ♀ 26"-29" ♂
24" ♀ 26" ♂
18" MIN.
24" MIN.
27"-28" ♀ 28"-29½" ♂
29" ♀ 30" ♂
32"-38" 36" STD FOR VISE & PLANING
34" MAX, 36" STD.
18" MAX
16" MIN. STOOL CLEAR.
1'-6"
6" MIN.
42" TO ₵
6"-8" BAR RAIL
BAR HEIGHT STD. 42"
39" ♀ 42" ♂
15'-10"

HIGHEST SHELF 72" ♂
HIGHEST SHELF 64½" ♀
FOOD PREPARATION 34"
FOOD PREPARATION 32"-34½"
PHONE DIAL ₵ STANDING 58"-60" ♂
PHONE DIAL ₵ STANDING 55" ♀
OFFICE AV. PHONE DIAL ₵ SITTING
OFFICE AV. PHONE DIAL ₵ SITTING 39½" ♀
SINK RIM 36" STD.
COOKING SURFACE 31½"-33½" ♀
21⅝" ♀ 38" ♂
26" ♀ 30" ♂
39" ♂
41'-3" OPT.
41" OPT.
17" ♀
16" ♂

60"-66"
78"
48" MIN.
24" MIN. 36" MAX
19"-17"
16"-18"   16"-18"
23"-43" SEE NOTE
54" MAX, 48" ADA
60" ₵
64" MIN, 74" ADA

NOTE: LAV HEIGHT VARIES
w/ USER HEIGHT:

USER HT	OPT. LAV. HT.
4'0"	27"
4'6"	31"
5'4"	37"
6'2"	43"

Small but hopeful steps are being made toward gathering the power of the sun more directly, and these first few attempts must surely begin a new science of enjoying the sun's beneficence, but the whole benefit of solar energy isn't possible until we turn a corner of time and we all experience a new renaissance of awareness, belonging, and caution. To have that renaissance we will be obliged to

Segen packs have two qualities that are usually antithetical: they are both beautiful and friendly. The construction is elegantly contrived and the materials — laminated hardwood, red latigo leather, #10 brown duck canvas, felt, and cord — are so natural and harmonious that they murmur and chuckle, wanting to be touched. Ed Segen maintains that his work is "advanced simplicity," a good phrase for a beautiful wedding of strong, traditional materials and impeccable design. You might be tempted to hang the Segen Pax on your living-room wall, but the pack itself would argue you out of it and get both of you out on the trail. Pricey, lovely, an aristocratic grandchild of Trapper Nelsons and other early wood frames and packboards.

Segen Pax
190 Rover Loop One
Eugene, Ore.   97404
(503) 689-8383

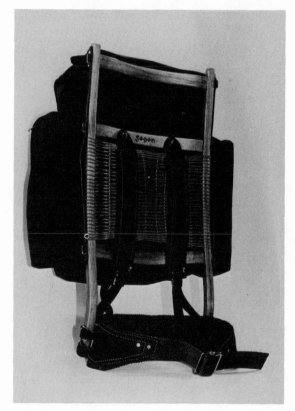

"Pak Carrier" by Joe Dodge and the White Mountain Hut Boys, U.S. Bobbin & Shuttle Co.

put away many of our attitudes about shelter, transportation, and living together. Some attitudes have been established recently, through a freewheeling time of cheap energy and expansion that cost us more than we knew. Some of the attitudes we will leave behind are older and subtler, and will be harder to let go. The wide sprawl of suburbia with its schools and markets and businesses accessible

Glue

by Jan Adkins

Glue is a vanity, of sorts, a check against the tide of entropy. We put together things and hope that they'll stand up to the rigors of use today and to time later. Some of us overbuild, cringing at all the blows and falls that are hiding in the years ahead. Some of us (often chair makers, for some reason), as experience and success build up, pare down the structure to a whisper over what we think is necessary, daring shocks and age, believing we have hickory and maple on our side. It all goes, eventually, but glue makes a difference: maybe glue can make it outlast us and spare us the embarrassment. That's enough to ask. Your great-grandchild can fix it . . . with more glue.

Unless you fix your expectations of glue it will be a disappointment. There are two ways to regard it: as a leveler, and as a liquid fixit nail. It is rather more than one and rather less than the other.

If you were to begin at the cornice of an old brick building you could lift the bricks off, course by course, with little or no resistance from the mortar, grown old and crumbly. Yet the mortar is still a major part of the building's strength: it levels each brick, keeps it in alignment with the rest of the structure, and provides a bearing surface for the entire brick. Wood is imperfect and changeable; handwork is imperfect and erratic; joints in wood are imperfect and uneven. It is possible and proper to view glue as a temporarily liquid substance that fills the imperfections of a joint, offers a whole bearing surface, and fixes the parts in a more stable relationship.

As a liquid fixit nail, glue is almost always a failure. Strength is a product of the whole structure and not of a structure's parts: this is the most basic tenet of design in woodworking, architecture, writing, or any other discipline. Two corollaries to this tenet are: never rely on your fastenings; and never tempt fate. Ideally a design uses fastenings (glue, nails, dowels, screws, etc.) only to hold pieces in positions of strength, and not as structural elements. Stress must be passed from one element of an assembly to another in the gentlest and most direct way possible. At joints the stresses build up and change direction. As the whole structure, say a table or a chair, bears weight, it flexes. Its angles and dimensions change slightly but powerfully, and this stress is concentrated at the joints. You could conceivably cobble up a

only by car is a recent pattern, and that can shift toward a tighter, less energy intensive pattern without a major change in attitude. The life of the Sun Age shouldn't be so grim, though. It will not be a sterile life in an eighty-story glass apartment building. As a matter of fact it will probably be more like the close, friendly life of an American town or village in 1880 — without the disease, shortages, isola-

chair by gluing together pieces of 1×4. You might even sit on it, several times, but you are relying on your fastenings, tempting fate, and you will be precipitously surprised.

If you can design a piece that will stand and function without glue, your confidence will multiply when you glue its parts together, perfecting imperfect joints, making sure of full bearing surfaces, and (behind fate's back) keeping it stuck together. If you can't design a piece that will stand without glue, you haven't tried hard enough. If you try and still need the glue, use it well.

Take care in this: wood is directional, and given its three aspects — side and face and end grains — the strength of the bond varies. Side grain to side grain makes an excellent bond under good conditions; this is fortunate because wide surfaces like tabletops and panels can be made up of side-bonded pieces. Tongue-and-groove joints, splined joints, and the like have been used to provide more surface area for gluing, but they have little advantage over a well-made plain joint. The difficulty of producing more complicated joints and the resulting inaccuracies often cancel out any advantage. Face grain to face grain can be a good joint, given care. Side grain to face grain is, similarly, a good combination. End grain to end grain is hopeless; because of the porosity of wood you cannot rely on any bond with end grain, even when it joins face or side grain. Further, internal stresses are inherent in such joints because of the relatively great difference in the way moisture changes the dimensions. For these unreliable combinations

Donna Meilach	$5.95
Creating Small Wood Objects	248 pp.
New York: Crown, 1976	Illustrated

A picture book of examples with half a hundred pages of good information on wood and various techniques, this book provides an ideal selection of objects for a craftsperson about to start work: enough beautiful pieces to inspire you and enough schlock to make you realize that you can do better than *that*.

Herbert L. Edlin	$12.50
Woodland Crafts in Britain	182 pp.
Newton Abbot, England: David and	Illustrated
Charles, Ltd., 1949, second edition 1973	

I had expected this to be a dry tome, but the old professions and the old professionals came alive. I enjoyed this as much as any of the books I've reviewed here. It is full of lore and information and technique, some of use in your workshop. This book is a store of knowledge from another time, a treasure.

George Jack	$8.95
Woodcarving: Design and Workmanship	296 pp.
London: Pittman, 1903, reprint 1978	Illustrated

This recent British reprint of a 1903 text contains line drawings, photos of examples from British cathedrals, and the feeling — still intact — that your wonderfully skilled grandfather is teaching you the trade. The instruction is still valid, and when Mr. Jack starts to talk about the "obstinacy of the wood fibre" you would do well to listen.

Aldren A. Watson	$7.50
Country Furniture	274 pp.
New York: Thomas Crowell, 1974	Illustrated

There is considerable knowledge here, a measure of history, and a display of skill . . . at illustrating and woodwork. Watson's progressive drawings are instructive, his text and graphic explanations clear, his glossary useful. I

tion and uncertainties of those slower times. The way we think of shelter, our ingrained and almost animal sense of what makes a place to eat and sleep, what makes us most comfortable — these are feelings that reach far back to the bogs and the forests where we began, and they must be accounted for.

Those old feelings about shelter are intertwined with our feelings for wood like the in-

like this book and think it would be a good addition to a woodworker's library.

Drew Langsner $9.95
Country Woodcraft: A Handbook of 304 pp.
Traditional Woodworking Techniques and Illustrated
Projects
Emmaus, Pa.: Rodale Press, 1978

This is one of the best books on handwork with wood, the result of a body of skill displayed in a harmonious composition of text, design, photos, and illustrations, all of a quality. Recommended.

Michael Dunbar $12.50
Antique Woodworking Tools 192 pp.
New York: Hastings House, 1977 Illustrated

Windsor Chairmaking $9.95
New York: Hastings House, 1976 153 pp.
  Illustrated

It is obvious that Michael Dunbar knows what he's talking about, and though I may disagree with his philosophy I appreciate his mastery of technique. He is a hand-tool craftsman of rare knowledge and ability and he holds sway at the Strawberry Banke historical museum shop in Portsmouth, New Hampshire. His books praise hand tools to the exclusion of all power tools, and the exclusion is not gentle. In his harangues against voltage he reminds me of that grand old sailmaster L. Francis Herieshoff, who believed that any man aboard a powerboat was a lazy dullard, not to be trusted, and the women aboard were mostly harlots. I can disagree while finding such open and unbuttressed belief refreshing. Dunbar's case would be stronger if it sounded less like a self-advertisement and if it were edited with a sharper and more willing blue pencil; he tends to overstate and to pile up his clichés. There are raisins in this sticky pudding, though, enough to make a tidy pile of solid knowledge and to justify two useful books. I could wish for more and better photos in *Antique Woodworking Tools* and for some good illustrations. *Windsor Chairmaking* has more photographs and addresses its subject completely and well.

tertwining grain of an elm, and just as hard to split. For most of us the thought of a sheltered place, warm and dry and safe and calm, has a woody glow: the tight-edge grain of strong oak beams above us and its face-grain figure in paneling around, the solid joinery of well-made chairs in walnut or beech, logs crackling on the grate. The shape and feel of wooden things are a part of our aesthetic, the

the ingenuity of the woodbutcher has provided alternatives, usually cuts that interlock the pieces in at least one plane and expose other than end grain to gluing.

Beyond the basic structure of wood, other conditions affect a bond: the species and type of wood, the moisture in the wood, the condition of the surfaces, the fit of the joint and the pressure of bonding, and the assembly time.

The density of the wood is a factor: dense woods need more care in gluing than do light woods. Expect fir to glue more easily than walnut, and expect sapwood to be easier than heartwood.

Some woods — lemonwood, teak, yew, and cedar, for instance — are "oily" and resist good bonding.

Moisture affects the strength of a bond in several ways. Wet wood will absorb little glue and the eventual shrinkage will put stress on the bond. Overly dry wood will absorb the glue too quickly and result in a *dried joint* and a weak bond. Gluing adds moisture to the wood locally; the best bond would result from joining parts whose moisture content was brought to its working level (the average moisture level it will experience) by the addition of the glue's water, but such a distinction is a pretty fussy matter outside a laboratory or a humidity-controlled factory.

This is a good place to note the effect of working wood too soon after gluing. Moisture swells wood; if you sand or plane a joint before swelling (due to the glue's moisture) is gone, you take away a greater amount of wood at the joint, and when the wood does dry there

will be a noticeable depression or groove at the joint. At room temperature, about seven days should see the normalization of moisture level; at 160°F, about twenty-four hours.

The surfaces of the joint should be as perfect as you can make them: clean, true, smooth, and dust-free. It was once customary to rough gluing surfaces with sandpaper, rasps, or special toothed planes, but well-run tests suggest no advantage and, indeed, the possibility that dust held in the striations may weaken the bond.

The film of glue between pieces must be thin and contiguous, and should exclude air from the joint. The two agents that can achieve such an ideal are the fit of the joint itself, as close and true as possible, and clamping pressure, which pushes the pieces together even more. The viscosity of the glue used also affects the film condition. Dense woods bond best with thick glues, soft woods with thin glues. Thicker glues want great clamping pressure to achieve the ideal film; thin glues can do with light pressure. Clamping should be maintained long enough for the glue to develop strength to resist interior stresses.

Assembly time is the period between spreading the glue and assembling the joint. The glue should set up to a tacky state before assembly, but there is a danger in waiting too long: a *chilled* or *dried* joint. A dense wood, being less absorptive, needs a longer assembly time and you will want to know the glue's *pot life*, a span of time in which the glue is usable after mixing or out of the bottle. Assembly time varies, then, with absorption and so

very way we perceive things. Why is this wood fixation so close to the bone?

The first tools we used to increase our strength or reach were certainly wooden clubs and staves for general hitting, levers for lifting, and pointed sticks for grubbing the earth or goring bison. Wood was at hand, replaceable, and workable. Ogg and Sons, rudimentary carpenters, had forests of materials from

end grain must be given special consideration: a light film of glue spread on end grain some time before the final spread will seal it and insure sufficient surface glue.

There are dozens of adhesives to bond everything, from skin to sailcloth to metal. Contact adhesives, epoxies, instant glues, rubber cements, sealants, and mastics are used by builders and cabinetmakers for special jobs, but the woodworker is concerned primarily with six or seven adhesives: hide glue, white glues (polyvinyl acetates), aliphatic resins, urea formaldehydes, resorcinols, and casein glues. Other urea resins are commonly used in furniture factories but require setting with high-frequency heat tools and are seldom part of an individual woodworker's shop.

Hide Glue

Form:	Thick fluid in squeeze bottles, or flakes to be mixed with water and heated
Advantages:	High, stable strength, it provides ease of use and good gap-filling properties. Instrument makers use hide glue because the bond can be broken for repairs, without damage to the parts, by inserting a hot palette knife into the joint.
Disadvantages:	Its yellow-brown color makes joints apparent in light wood. It loses strength with moisture or heat and can be brittle, failing under shock.
Strength:	High
Water resistance:	Low
Gap filling:	Excellent
Setting time:	8 hours at room temperature
Cleanup:	Warm water
Brands:	Franklin Liquid Hide Glue

White Glue (polyvinyl acetate)

Form:	A white, viscous fluid in a squeeze bottle or jars
Advantages:	Readily available, this widely applicable adhesive is easy to use, easy to clean up, and dries clear.
Disadvantages:	It is unsuitable for use in conditions of heat or moisture, and too soft to sand smooth.
Strength:	Moderate
Water resistance:	Poor
Gap filling:	Good

which to choose staves of any thickness, clubs with just the right heft and wallop. The material was soft enough to shape with fire and scrapers, light to carry and throw, and it was strong. Building shelters, Ogg and Sons had only one material that could span useful distances, hold the weight of a tent's hides and the wind loads on them, support thatch roofing over a permanent house, or bear a wooden

Setting time:	30–60 minutes drying, 8 hours curing
Cleanup:	Warm water
Brands:	Devcon Grip-Wood Duratite (Dap) Elmer's Arts & Crafts Elmer's Glue-All Elmer's School Glue Wilhold Glu-Bird White Glue

Aliphatic Resin

Form:	A light buff viscous fluid in squeeze bottles or jars
Advantages:	A very strong woodworking glue, it dries clear and hard enough to sand smooth. With quick tack, short clamping time, it's easy to use and to clean up.
Disadvantages:	Under constant loading this otherwise strong glue will allow parts to creep out of place unless they are properly supported by adjoining structure.
Strength:	High
Water resistance:	Low
Gap filling:	Good

Setting time:	30 minutes clamping, 8 hours curing at room temperature
Cleanup:	Warm water
Brands:	Elmer's Professional Carpenter's Wood Glue Franklin Titebond Glue Wilhold Glu-Saver Aliphatic Resin Glue

Urea Formaldehyde

Form:	A gray powder to be mixed with water
Advantages:	A high strength, nonstaining, water-resistant glue, it makes an invisible joint.
Disadvantages:	It must be mixed, then used within a few hours. It's a poor gap filler, and requires long clamping time.
Strength:	High
Water resistance:	High
Gap filling:	Poor
Setting time:	9–13 hours at room temperature
Cleanup:	Soapy warm water
Brands:	Elmer's Plastic Resin Cascamite Glue

house and its family above a bog. Stone could span a distance but at the cost of formidable labor quarrying and transporting, a heftier foundation and supports, and the strain of lifting it into place. Wood could be handled, shaped, and fastened.

Its strength lay in its lamination of dense winter wood and light spring growth. Even Ogg and Sons could see that those growth

Weldwood Plastic Resin Glue
Wilhold Marine Grade Plastic Resin Glue

Resorcinol

Form:	Two parts, liquid and powder
Advantages:	A very high strength and heat resistant adhesive, it is absolutely waterproof.
Disadvantages:	Difficulty of mixing, short pot life (3–4 hours), long clamping time and a dark red color that stains joints make this a glue primarily for marine and exterior specialties.
Strength:	High
Water resistance:	Excellent
Gap filling:	Good
Setting time:	Varies with temperature; clamp and cure in 6–10 hours at room temperature (70–80°F)
Cleanup:	Sanding, solvent suggested by manufacturer
Brands:	Elmer's Waterproof Glue

Casein

Form:	Dry, white powder to be mixed with water
Advantages:	The principle advantage casein holds over all other adhesives is that it will set at any temperature over freezing with moderately high strength and fair moisture resistance. It is also excellent for oily woods (lemon, yew, teak, cedar).
Disadvantages:	Difficult to mix, casein glues stain some woods and are abrasive to tools.
Strength:	High
Water resistance:	Fair
Gap filling:	Good
Setting time:	2–3 hours clamping, curing dependent on temperature

rings gave it a directional strength they had to deal with as they split, shaped, and used their handiwork. It was unyielding in compression against the end grain, springy against the face grain, stiff against the edge grain. Wood was the strongest material they had in the shop . . . or anywhere. Stone was harder but heavier; bone was lighter but more brittle. Per unit of weight, wood was the

Eric Sloan $7.50
American Yesterday
New York: Funk and Wagnalls, 1956

Each of Eric Sloan's books on earlier America is really part of one serial book extolling lost beauties and forgotten virtues. Mr. Sloan's peppery personality may have something to do with his impatience at the necessity of living in this century, but his pepper is a pungent spice that delights us. We may maintain that our forefathers were every bit the silly, greedy, neglectful beggars their sons and daughters became, but we must also agree that the glory of the American landscape has faded since their time, and Sloan's mannered but controlled drawings bring some of its loss to us. His style of drawing and his knack of showing process with pictures and captions have fired imaginations for many years and have started more than one young illustrator toward finding his own style of explication with pictures and arrows and words. This is as good a place as any to acknowledge my debt to him, and I do so freely and thankfully.

Sloan books are part of my ideal bathroom library because their illustrations are so informative and often so much fun that reading the text is almost forgotten — along with such rigors of those early days as smallpox or rickets or the prize hog eating the new baby brother. He provides a fine resource for hand-tool usage and his explanations of early barns and bridges make the run of his work almost a textbook for first-year architects.

American Yesterday   (New York: Funk and Wagnalls, 1956). $7.50.
Reverence for Wood   (New York: Funk and Wagnalls, 1965). $9.95.
Diary of an Early American Boy   (New York: Funk and Wagnalls, 1962). $7.95.
American Barns and Covered Bridges   (New York: Funk and Wagnalls, 1954). $7.50.
The Seasons of America Past   (New York: Funk and Wagnalls, 1958). $7.50.
A Museum of Early American Tools   (New York: Funk and Wagnalls, 1973). $7.95.
Age of Barns   (New York: Funk and Wagnalls, 1966). $15.95.

Our Vanishing Landscape   (New York: Funk and Wagnalls, 1975). $7.50.

Byron D. Halsted $6.95
Barns, Sheds and Outbuildings: 240 pp.
Placement, Design and Construction   Illustrated
Brattleboro, Vt.: Stephen Greene Press,
1977

"Barns can be pleasing objects, and impart an impression of comfort and completeness upon all who see them." Indeed. This is a "near-facsimile" reprint of an illustrated treatise first printed in 1881, written by one Orange Judd, agricultural correspondent for the younger *New York Times*. There is a great deal more than whimsey and enjoyment here: there is practical information, sound design, and real wisdom. One of the most difficult projects for second-year architects under Professor George Tilley at Ohio State University was the design of an outhouse for a country grade school. We cursed and mumbled; we protested. Didn't he realize this was 1963? We learned that

A treadle lathe

Fine Woodworking $14/year, $3.00/issue
Newtown, Conn.: Taunton Press Bimonthly

Biennial Design Book $10.00

Only one magazine makes any claim to addressing woodworking's inner game. The others are blatantly do-it-yourself vehicles for power-tool adverts. *Fine Woodworking* is the *Scientific American* of wood skills, a refined and restrained periodical that carries articles on the minutiae of thumb gauges and tambour cutting, on the distant reaches of wooden clock movements and Gothic tracery, and on the grosser carpentry of building barns with green wood and hanging a door that shuts properly. Well photographed and illustrated, each issue is an invitational workshop of techniques and a gallery showing of work by the best. Their *Biennial Design Book* is stunning, a display of talent to make you understand why wood chips smell so fine.

architecture was a discipline of problem solving with techniques appropriate to the place and time, and the given sewage setup for this out-county school gave us a rough two weeks. This book could well be a textbook for architecture students, and if they digested the hard-learned, time-grown, low-tech lessons in it, their own designs could be simpler, more logical, and could impart an impression of comfort and completeness to all who saw them. A very good investment for anyone intending to build a house, arrange a farm, landscape an estate, or put up a toolshed.

The Craftsman in America $5.75
Washington, D.C.: National 200 pp.
Geographic, 1975 Color photos

If you work with your hands, the roots of your craft are pictured here in splendid color photos. The wood section is a delight, the overall design of the book is beautiful, and the coverage of other crafts proves that all skilled hands are working at the same varicolored tapestry.

strongest, most convenient, most widely applicable material available.

It still is.

The qualities that made wood so attractive to Ogg and Sons make wood attractive to us, especially for shelter. Wood is perennially available, a renewable resource for as long as sun and rain and soil survive. It is available for less energy cost than any other building

Log house in late-19th-century Montana, with occupants

"LOG HOMES"

It's not that I am against logs, but they lack a great deal in the way of an ideal building medium. Their insulation value is not particularly high; they are neither uniform nor dimensionally stable; they invite decay and attack by insects and, for all I know, beavers.

There are some hearty recommendations in favor of log building outside the megalopolis, though. If you are unconcerned with the additional expense of labor (if you are a homesteader or an eccentric duke, say), then the initial cost is low. Felling logs and using them *in situ* would seem to be a savings in overall energy, considering the trip to the sawmill, sawyering, the waste wood, and trucking the dimensioned lumber back to the site. There is a beguiling continuity in clearing land from the forest and building shelter from the trees you take down, and in the tie with the older settlers and their log homes. There is an interest

in log building that defies the lateness of the century and there may be some merit in it.

B. Allen Mackie, a Canadian builder of log homes and dean of a well-enrolled school of log building, insists that the term "log cabin" is an Americanism and a misnomer with its roots in the slave quarter cabins. The "lawg caybun" philosophy of the "rustic" is what Mackie objects to and it may well be the reason I, among many, have a building prejudice that excludes logs and moss chink. The best examples in the three best books we have, however, show a sophistication and a fine turn of detail I had not expected. If you have the time and the trees and a good wood stove, log building may be a livable alternative to balloon framing.

Of the log-building books we've read we recommend starting with William C. Leitch's *Hand Hewn*, an appre-

medium: wood requires only cutting, hauling, milling, and seasoning. Other media must be mined, crushed, smelted, rolled, and transported — a huge investment of energy. The earth smelts wood in the furance of the sun. Time and the elements process it.

Through the Second World War the heart of London was Churchill's bunker, shored and protected by ancient oak timbers, ribs from

ciation of the log home in words and photographs with a good bibliography, and E. Allen Mackie's *Building with Logs*, a crude but dense compilation of real skills and needs.

Garden Way Publishers have an investment in the splinter cultures of homesteading, subsistence farming, and escapism. They are wonderful, inventive, right-minded folks, and I wish they would make more of an investment in style and competent illustration and more watchful editing. Their *Build Your Own Low-Cost Log Home*, by Roger Hard, seems to be, illustrations notwithstanding, a workable treatment. I had dismissed W. Ben Hunt's *How to Build and Furnish a Log Cabin* as embarrassingly rustic, but it does cover some special techniques.

Dan Beard's *Shelters, Shacks, and Shanties*, in paperback reprint from the original 1914 edition, is time-travel entertainment, back to your grandpa's youth, a recommended bauble and something that may wow and intimidate a fourth- or fifth-generation space child.

<table>
<tr><td>

William C. Leitch
Hand-Hewn
San Francisco: Chronicle Books, 1976

</td><td>

$4.95
120 pp.
Illustrated

</td></tr>
</table>

E. Allen Mackie
Building with Logs (1972)   76 pp.
Box 1205, Prince George, B.C. V2L4V3   Illustrated

Roger Hard $6.95
Build Your Own Low-Cost Log Home   198 pp.
Charlotte, Vt.: Garden Way Publishing,   Illustrated
1977

W. Ben Hunt — 38 $3.95
How to Build and Furnish a Log Cabin   166 pp.
New York: Collier Books, 1974   Illustrated

D. C. Beard $2.75
Shelters, Shacks, and Shanties   243 pp.
New York: Scribner's Sons, 1914,   Illustrated
republished 1972

one of Lord Nelson's flagships. Wood is strong, many times stronger than steel for its weight. Its strength is not the powdery intractability of masonry but a pliant strength that responds to weather and shock without yielding, a quality especially appropriate to earthquake areas.

A surprising fact is that timber construction can be safer and more lasting against fire than steel. Wood is an insulator; it passes heat

WHY IS WATER COMING FROM THE LINEN CLOSET?

If you own a house, or plan to, you should have a reference against all the dismaying accidents that time, decay, vanity, and ambition will foist off on you. You must know how to repair damage from insects, children, and other pests. You must have an acquaintance with plumbing and wiring (first-year physics books say the two are very similar but I don't believe them). You must be able to paint and to remove paint, to put up and take down walls, to shingle, and to fix leaks. You can avoid all responsibility with enough extra money for plumbers, electricians, roofers, etc., but you will need patience with the gold, because the contemporary skilled laborer doesn't rush to your door at a phone call. The obvious answer is to have detailed, reassuring books that tell you what to do. I see it as a necessity, and there are at least two fine choices.

My criteria for such books are breadth, completeness, and encouragement. Both the single-volume *Reader's Digest Complete Do-It-Yourself Manual* and the *Time-Life Home Repair and Improvement* ten-book set rate very high in breadth of subjects covered and in detail of answers, advice, and information. Because the Time-Life series is more attractive, its illustrations and photographs better and larger, its pace more leisurely, and its approach more personal, I give it the edge for encouragement (the hell-yes-I-can-do-that factor). If you are more secure than I (and the odds are in your favor) the Reader's Digest book is a busy but powerful reference and a better buy.

Reader's Digest $17.95
Complete Do-It-Yourself Manual   600 pp.
New York: Norton, 1973   Illustrated

Time-Life Home Repair Series $7.95 each
Paint and Wallpaper   128 pp.
Space and Storage   Illustrated
Basic Wiring
Weatherproofing
Heating and Cooling
Plumbing
New Living Spaces

poorly, even to itself. Steel conducts the heat of a fire quickly and has lost any structural integrity at 1450° F. A heavy timber can endure the extreme temperatures of a fire (1700° F. and above) long after heavy steel has buckled and collapsed. The charred wood on the outside surface of a burned timber actually insulates it further. A redwood firewall will stop or at least delay a fire better than a steel

Kitchens and Bathrooms
Masonry
Roofs and Siding
New York: Time-Life Pub. Co., 1976

Harry Walton $3.95
Home and Workshop Guide to Sharpening   160 pp.
New York: Harper and Row, 1967, second   Illustrated
edition 1976

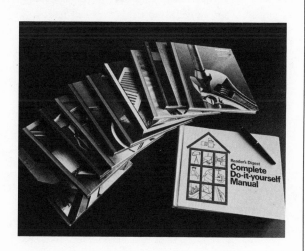

You'd better get this book. It shows you how to sharpen everything but wit, usually more than one way. With its breadth of treatment, its explicit photos and diagrams, it is a shop reference you will not want to be without.

Dermot McGuigan $4.95
Harnessing Water Power for Home Energy   102 pp.
Charlotte, Vt.: Garden Way Publishing,   Illustrated
1978

Dermot McGuigan $4.95
Harnessing the Wind for Home Energy   124 pp.
Charlotte, Vt.: Garden Way Publishing,   Illustrated
1978

If we talk about energy-efficient houses it seems appropriate to talk about energy. It is possible to pare energy needs down to the scale of wind generation or water generation, if you have the terrain and weather for one or both. It may be necessary, more than possible, one day. These two small books are factual, hopeful, serious, well mounted, and, perhaps, prophetic. The word "decentralization" occurs more frequently in technical journals and is about to move into the public idiom, helped by efforts like these small books.

Dale L. Nish $7.95
Creative Woodturning   248 pp.
Provo, Utah: Brigham Young University   Illustrated
Press, 1975

A complete course in turning through detailed photographs, this book takes you from primary theory through abstruse technique, and includes examples and ideas, substantial information, and straightforward design. A recommended book.

wall, which simply passes the heat and ignites material on the other side. A wooden fire escape is usable when its steel counterpart would be glowing and structurally ruined. Dr. William Harlow, in his excellent *Inside Wood,* details a plywood strongbox to protect papers that has withstood a full hour at 1600° F. to 1740° F. without damage to its contents. In a metal box the papers would have been incin-

"GO HOME"

I am an improper reviewer of books about house building since I often toy with the belief that stiff fines and perhaps even sentences of confinement should be levied on anyone building a conventional house at this late date. It soothes my outrage only a little to reflect that whoever does will be fined by the power company, confined by boxlike tradition, and will pay through the snout for the privilege. To be strident about solar — or at least energy-conscious — houses may be curmudgeonly. If the cute capes and the ranchburgers weren't wasting energy we're all going to need, it would be merely foolish not to get in touch with the sun, and everyone should have the right to be foolish, but we'll all pay for their folly down the road a piece.

If you plan to build a home, there are books that will help more than your understanding of stud framing. Begin with *30 Energy Efficient Houses You Can Build* which will show you working examples of intelligent design and proper expediency. The book is a bargain for the plans, ideas, advice, and photos. It will provide a smorgasbord sampler of appropriate building technology.

From the Ground Up provides a capsule course in architecture that, taken with a fresh mind and a serious involvement, may be as good as the technical side of the first two years of architecture school in quite a few ivy-bound institutions. It deals with human needs and living patterns, with floor loads and wind loads, and with construction details and methods. Highest recommendation.

Designing and Building a Solar House has much of the technical information you need to plan your solar house in an agreeable and comprehensive presentation, understandable charts and diagrams, and prime examples.

Standard reference in the trade for numbers and documented concepts has been *Solar Energy.*

The Complete Greenhouse Book deals with more than plant environments, but if you've ever seen the beautiful Cape Cod or Prince Edward Arks of the New Alchemy Institute a greenhouse may be enough for you.

So You Want to Build a House is a more traditional approach to building, full of sound facts and revelations but unenlightened in a solar sense. (I am such a bigot.)

Illustrated Housebuilding is too childlike and funky for my jaded tastes, but *Your Engineered House* is iconoclastic enough to clear a raft of preconceptions out of your head.

Wood Houses for Country Living is a rehash of a standard U.S. Department of Agriculture reference *Wood Frame House Construction,* by the same author. Check the prices.

Garden Way's *547 Easy Ways to Save Energy in Your Home* is hardly worth the effort, but their *Low Cost Pole Building Construction* details an overlooked, logical, fast, and economical construction method.

I am confused about timber frame construction. *The Timber Framing Book* is a fine piece of work, complete and charming, but the charms of timber framing still seem to be an anachronism. Balloon framing and hem-fir studding don't make nearly as nice a dollhouse as mortise-and-tenon maple framing, but both get covered by sheathing, insulation, and interior cover and where is your furniture-crafted joint then? The form and the pattern of life in a house are the cardinal proofs of its worth and, if you are not convinced that heavy-timber framing will last longer (I am not; I trust plywood) a faster and cheaper building method may be more adaptable and less precious than the old ways of barn construction. If it appeals to you without nostalgic disguise, then *The Timber Framing Book* is a good treatment.

Two very personal approaches to house building are in *The Green Wood House* and *I Built Myself a House.* The author of the former is an experienced and well-spoken builder of low-cost homes. His premise, that rough-sawn green wood can be used to advantage in residence construction with some changes in details and technique, is convincing. His details and methods are respectable and his lack of gypsum board cosmetics is refreshing. *I Built Myself a House* is a personal statement rather than a manual. I would not count on it for more than glib encouragement, because it smacks of the I-don't-know-much-about-floor-loads-but-I-know-what-I-like school.

If you are going to fix up an old house, and if you are not intimately acquainted with casements and bird's-mouths and cant strips and conduits, *So You Want to Fix Up an Old House* is a friend and adviser and a tonic of confidence.

erated. An insulated firebox affording equal protection would cost two to four times as much as the wood box.

Wood's endurance is not limited to fire-resistance. Time, the lord of decay, is kind to well-built wood structures. Some of our oldest buildings are wooden. Japan's Golden Hall was built in A.D. 679, survived a major fire in 1949, and is still open to the public. Scandi-

Alex Wade and Neal Ewenstein
30 Energy Efficient Houses You Can Build
Emmaus, Pa.: Rodale Press, 1977
$8.95
316 pp.
Illustrated

John N. Cole and Charles Wing
From the Ground Up
Boston: Atlantic Monthly Press/Little, Brown, 1976
$7.95
244 pp.
Illustrated

Donald Watson
Designing and Building a Solar House
Charlotte, Vt.: Garden Way Publishing, 1977
$8.95
282 pp.
Illustrated

Bruce Anderson
Solar Energy
New York: McGraw-Hill, 1977
$23.75
374 pp.
Illustrated

Peter Clegg and Derry Watkins
The Complete Greenhouse Book
Charlotte, Vt.: Garden Way Publishing, 1978
$8.95
280 pp.
Illustrated

Peter Hotton
So You Want to Build a House
Boston: Little, Brown, 1976
$6.95
232 pp.
Illustrated

Graham Blackburn
Illustrated Housebuilding
Woodstock, N.Y.: Overlook Press, 1974
$4.95
155 pp.
Illustrated

Rex Roberts
Your Engineered House
New York: M. Evans and Co., 1974
$4.95
Illustrated

L. O. Anderson
Wood Houses for Country Living
New York: Drake, 1977
$5.95
108 pp.
Illustrated

L. O. Anderson
Wood Frame House Construction
Washington, D.C.: U.S. Dept. of Agriculture, Forestry Service, 1970
$2.60
223 pp.
Illustrated

Roger Albright
547 Easy Ways to Save Energy in Your Home
Charlotte, Vt.: Garden Way Publishing, 1978
$4.95
124 pp.
Illustrated

Doug Merrilees and Evelyn Loveday
Low Cost Pole Building Construction
Charlotte, Vt.: Garden Way Publishing, 1979
$5.95
102 pp.

Stewart Elliott and Eugenie Wallace
The Timber Framing Book
Kittery Point, Me.: Housesmith's Press, 1977
$9.95
169 pp.
Illustrated

Larry Michael Hackenberg
The Green Wood House
Charlottesville, Va.: University Press of Virginia, 1976
$4.98
144 pp.
Illustrated

Helen Garvey
I Built Myself a House
San Francisco: Shire Press, 1977, revised 1979
$2.50
128 pp.
Illustrated

Peter Hotton
So You Want to Fix Up an Old House
Boston: Little, Brown, 1980
$7.95
320 pp.
Illustrated

navian log halls and dwellings from the year 1000 are still robust and sound.

Praising wood is not enough. Seeing its faults you may discover overlooked virtues and, more important, a whole gray area between vice and virtue that constitutes craftsmanship. Wood suffers from its virtues: it is organic, a living material, each tree unique and beautiful of itself.

"FRESH REFERENCE"

There is an encouraging trend in publishing toward small, paperbound books of good photography. There is a need for them: to leaf through pages of good design refreshes and grounds a worker, reacquaints a designer with the high standards of the art, and promises the craftsman that past the tedium that is a part of every skill there are rewards.

These four books are full of excitement, a commodity rare enough to remark, of very good photographs, good designs, inspiration, and good cheer. A craftsperson's library wants much more than technical reference.

Jo Reid and John Peck	$5.95
Stove Book	107 pp.
New York: St. Martin's Press, 1977	Color photographs

Herbert H. Wise and Jeffery Weiss	$6.95
Living Places	80 pp.
New York: Quick Fox, 1976	Color photographs

Herbert H. Wise and Jeffery Weiss	$5.95
Made with Oak	80 pp.
New York: Linx Books, 1975	Color photographs

Ben Dennis and Bessie Case	$9.95
Houseboat	64 pp.
Seattle: Smuggler's Cove Publishing, 1977	Color photographs

As an organic material it can rot and decay. Some woods, like balsa, will rot away in a few months if not cut and dried; birch firewood will begin to rot from within if not split and stored dry. If the ground level around a structure changes with landscaping, with silt buildup, with brush or a dung heap against a barn wall, then hemlock sills, oak and maple posts, pine sheathing, and fir trim will decay,

In Search of the Soul of the Wooden Boat

by Jon Wilson

Wooden boats remind me a lot about what we've forgotten — or perhaps never knew. With rare exception, their shapes and structures reveal the accumulated experiences of thousands of years. They have pleasing shapes, for the most part; the material itself practically demands it. As if the grace of the forest trees were bequeathed in abundance to every plank sawn. And each plank, in turn, has carried with it the duty to lie gracefully in place, resisting to the end any move toward the awkward and angular. That duty was once well understood. Designers, builders, and just plain lovers of boats could respond in awe to the nature of wood and let their hearts and hands be guided by it.

Today we see too little. Our yearnings for more seem nostalgic, romantic, and unreasonable. But it has little to do with the past, more to do with the quality of the current creative spirit. At one time, no boats were built except by hand and eye alone. Experience and intuition guided the shapes of craft. The Viking and Polynesian traditions, almost literally poles apart in their needs and resources, contain remarkably unique craft, whose structure and design were governed largely by the resources

at hand. It was imperative to their success that they understood the limitations and the possibilities of the woods with which they worked, and in their understanding they created shapes that best enhanced those elements. Not only did that creative synthesis work well for its time, but the shapes thus created — the traditional craft of those seafaring nations — still leave us in awe.

When I look around at the stock plastic boats that are currently on the market, I wonder whether they leave anyone in awe *now*, much less ten years from now. But they serve their purpose — it just happens not to be *my* purpose.

So, what do the traditions and the future of wooden boats represent for us today? Certainly more than leisure-time diversions and delights, and certainly more than an annual series of headaches over chipped paint and open seams, greasy bilges and soggy berths. But I'm in the business of celebrating the existence of boats built of wood, and in affirming the future of that existence, so my focus may seem to be biased. I guess it is, for I truly believe that wooden boats have a lot to teach us about our purpose on the planet. (I also

lose their strength, and, in short order, fail. If the warm, moist air suspiring from a house is allowed to pass through the roof's insulation and meet the cold roof decking, it condenses as water droplets, commingles as streams running down the slope inside the roof, and collects on soffits and on the flat surfaces of the top plates to decay and weaken the structure. Any unprotected wooden part of a house that

believe that about music, houses, clear thought, sharing, and love.)

Picture this: A quiet man walks into an ancient wood to fell an old oak that stands with widespread branches. There's an old stand of cedar down in the swamp below, and when he's finished getting out the crooks from the limbs of the oak, and measured and cut that long, straight butt, he's going down to fell a couple of them.

His axe and his saw are clean and sharp, and he's been looking forward to this day for some time. The trees were marked last summer, perfectly suited to the building of a nineteen-footer, a sailing skiff whose model he carved out on a couple of June evenings and which sat on the mantle for months while he eyed her shape from every angle, living with her, as he knew he'd have to when she was done.

His work is purposeful and reverent — he moves through the wood with gratitude and care, delighting in this time with the very trees that will soon become lumber for the boat. He's in no hurry, for the lumber will be stacked for the winter and summer, drying slowly and comfortably — getting used to not being trees anymore, preparing for life as a boat.

On this crisp fall morning, he has felled and limbed up an oak and three cedars. The seeds of both have come down with the branches, and carefully he gathers them. The best will be sown in his infant woodlot; the rest will be scattered where the growth is thin. He believes in wood, and he knows its nature. He plants for the next generation, and the one beyond

Fenwick Williams 24-foot gaff-rigged yawl at the Brendze & Wester yard

that. Not to be remembered, because he won't be; but to give back to the earth, in return for her gifts.

If tomorrow's a good day, the neighbor will bring in his horses and skid out the logs. On the last trip the sledge will bring out the brush for kindling, and chunks for long fires. The

touches the earth is an avenue of opportunity for termites that consume the organic structure of a wooden structure. Even bringing firewood into a house can ferry wood-boring insects that spread and damage. Skill in designing and building and maintaining a structure protects the corruptible nature of wood: skill.

wood will be quiet again, with four fewer trees, four more stumps, a little more sunlight, and the green tips of a few cedar branches spread around.

The sawyer will come to pick up the logs and saw them with care for the boatwright. When he returns, his truck will be stacked high with lumber, and together they'll put it in the shed to dry.

The winter and summer will pass while the man waits patiently. The lumber will dry and shrink a bit as it yields the moisture from its inmost fibers. On the ends of a few cedar planks near the door, some small checks will occur — the draft has dried them too fast. But no matter, they're long enough still to get out good planks.

The mornings are frosty again when the time is right. With axe and saw, chisel and plane, he shapes the keel, stem, and sternpost. The carved model guides his hand in the certainty of experience and intuition. He trusts them implicitly — never falters on the way, and shapes the timbers with just the right balance of lightness and strength. The boat is, after all, to be agile and strong. His work is purposeful and unhurried. The upswept limb on the old oak is now the gentle sweep of the stem, and the straight-grained stock provides keel and sternpost. Bolted together, the backbone stands ready for frames and planks, and with battens and clamps her shape is defined. The stock for the frames is sawn out and made ready. Too stiff to be bent, it will lie in the steambox for an hour or so while the fibers are cooked just enough. Then, one at a time, they'll be rushed to the boat, clamped into place, and left to cool to their new shape. It's quick work, and easy, too. With enough moisture in the wood, and wet enough steam, you can tie a knot in a fresh-cooked oak frame.

The fragrance of new oak pervades the shop now, but tomorrow the shavings from new ce-

72

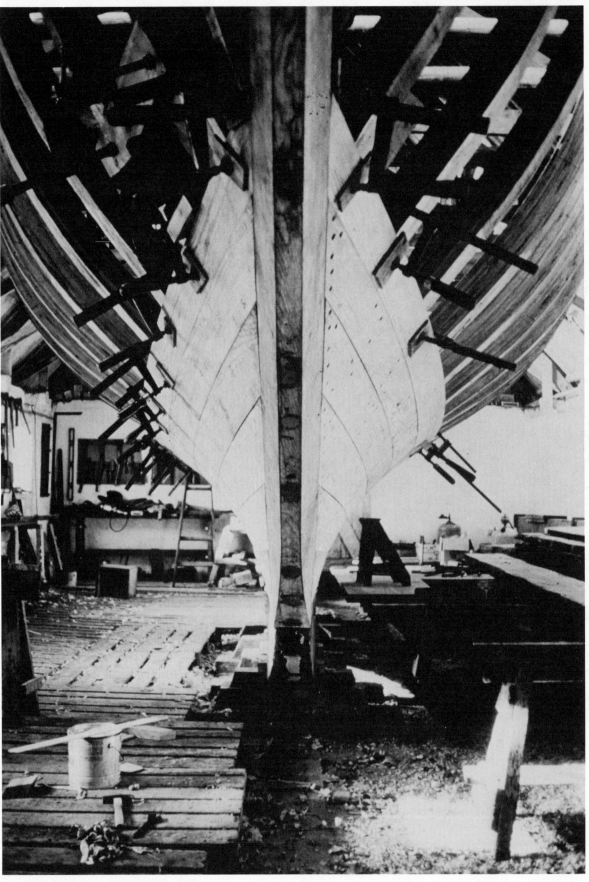

Colin Archer 30-foot cutter under construction in Denmark

Wood and water are an unstable pair, sometimes mating for an age, sometimes destroying their bond in a season. The city of Venice sits beautifully on pilings and has for centuries, but recent changes in the water level have exposed them to air for the first time since their sinking. Decay began immediately and the city is in danger. The timbers of a wooden ship live constantly with salt

dar planks will take over, their pungent aroma filling the place. And inside of two weeks the hull will be planked, as the idea takes form in reality.

Now the details will go a little more slowly as the boatwright considers his alternatives inside and on deck. Three thwarts? Four? Benches? Side decks? With a hull to step into, he can feel the subtleties of great and small changes, and doesn't begrudge the time spent in considering. When the questions are answered and alternatives chosen, he moves swiftly along. Paint for the surfaces easily worn, varnish for those that will show it off best, and oil for floorboards that must not be slick.

Solid spars shaped from a woodlot spruce will be oiled to give them a sensible finish. A sailmaker friend cuts and sews a cotton sail, and someone leaves a pair of oars rescued from a lifeboat no longer in service. The knowledge of a lifetime, and lifetimes before, has gone into this craft. She has grown from an idea into a living being, infused with the creative energy of the forest and work of a man. She has sprung from the earth's elements just as surely as the man who built her has. As the days grow longer and warmer, she waits with her builder, just as impatiently.

Then comes the day. The weather is warm, the tide is right, and she feels water beneath her for the first time. Afloat at last, this is the moment they've worked for, as the new sails fill and draw. Out of the harbor and across the great bay they sail, this man and his boat. Working together, they discover what she can

do, reaching, running, and climbing to windward. In the late afternoon the wind drops for a while, and before it picks up again he has sculled her home.

She is lively and responsive, with plenty of sail for the days with light breezes, and plenty to reef when the wind pipes up. What began as a vague thought has become a clear idea — a dream brought forth. She is lovingly modeled, lovingly built, and lovingly cared for. The hours of work are returned to him a hundred times over in pleasure. Rare is the day that he doesn't sail her, if for only an hour. And rare is the hour aboard that doesn't restore his spirit and affirm his love for the boat.

The soul of a wooden boat is an elusive thing. One can feel its presence in a derelict craft almost as much as in a well-loved one. But defining it is quite another thing. For me it is found in the blend of energies invested by the designer, the builder, and the owner. If all three investments have been made with love and care, the soul of the boat seems stronger.

There are very few things that we can dream into being, build with our own hands, and enjoy for a lifetime. Fewer still that we can buy from someone else. Wooden boats shine among them, in spite of some bad press in recent years.

It's said that wooden boats leak and rot, are a horror to maintain, and are too expensive to build anymore. Indeed, they're said to be dying out. Although most who own and love wooden boats must disagree, those who aren't sure why they're still attached to them are in grave danger of switching their allegiance.

water and are protected by it. The Swedish warship *Vasa,* raised in 1972 after a four-hundred-year sleep, is in excellent condition, though she must be constantly sprayed with water to prevent the air decay that is undermining Venice. Ships' timbers are, however, susceptible to the dry rot that sets in wherever fresh water, from condensation or rainwater,

Peterson schooner *Lille Dansker* under construction at the Hodgson yard in the late 1930s

lies in pools or between joints in the complex angles of a ship's scantling. End grain is especially vulnerable since it passes water along its length and rots timber, unseen, from within. At sea, as on land, borers attack wood: because of the tropical borer *teredo*, every vessel of size that plied the tropics in the age of sail was sheathed with tar, felt, thin white

Such people have often been the victims of inexperience and misinformation, or they may have discovered that the care of living things requires deeper commitment and responsibility than the care of a plastic facsimile does, and have neither the time, money, nor interest for such complication. Moreover, there are some very bad wooden boats in the world, examples of poor design, construction, and maintenance. The most lavish amount of love and care for such boats may at best only slow the process of deterioration. On the other hand are some wooden boats nearly a century old, which are still going strong for their owners.

The care of such fine creatures has become almost an occult science to some, perhaps because some of the elementary aspects of care and maintenance have been forgotten or obscured among the promotional boasts for today's miracle products — best suited to today's miracle boats.

Well, what about the rest of us? Perhaps the key to understanding lies in the continuing recognition of the wooden boat as a living thing; recognizing surface problems as symptomatic of deeper problems. Sometimes simple, sometimes not, all they require is attention. For instance, there's an area on the topsides of the hull where the paint blisters and peels every year, creating the otherwise unneeded work of having to scrape, sand, and spot with paint before doing the entire hull. One owner will find it a frustrating and unavoidable part of the ownership of a wooden boat, while another will explore the problem methodically. To begin with, it is undoubtedly a case of ex-

cessive moisture in the wood. Since wooden hulls need to evaporate moisture from the planking into the interior, it's very important that the structure be well known to good designers. But it wasn't always so. There have been many boats built whose ceiling (inner hull skin) is fitted tight from cabin sole to sheer clamp with nary a gap in it. Such a structure looks neat, adds stiffness, but prevents good ventilation, and without that, the wood becomes too saturated, and paint won't adhere. More often, the problem is a hanging locker, or a sealed-up compartment inside, which was left unventilated through oversight or carelessness. Cutting some openings so that air can circulate through the spaces between the frames will promptly allow the planking to breathe, and soon paint will adhere again. Trapped moisture — vapor or water — is the main problem in wood boats. In the right quantities it can peel paint and varnish, make a perfect place for rot, cover the interior with mildew, and turn rich mahogany to a spongy black mass. The ways to trap it are poor joinery, inexperience with structures at sea, total dependency on weird miracle substances, and inattentive or inexperienced maintenance.

The uninformed believe that the wood itself is at fault, as if by some malevolent decree, the wooden boats of the world have it in for their owners. Not so; they only need love and care, which for some is too high a price.

Another common woe is the chronic leaker. The boat that the owner is afraid to leave for more than a week, because the water will be over the floorboards or cabin sole. Or the boat

THE W.E.S.T. SYSTEM

Wood floats, and this reassuring quality is doubtless one of the reasons the earliest boats were built with it. But wood and water are not comfortable neighbors. Wood is hygroscopic: it takes and releases moisture, expanding and contracting in the process. You can toss your Chippendale highboy into the pond and it will float, but it won't stay together. Over centuries shipwrights have developed construction details that minimize the effects of wood's active discomfort, methods and materials that keep vessels tight against expansion, contraction, rot, and attack by borers. Some difficulties couldn't be overcome: such as fastenings, which tend to loosen; soaked wood loses its strength with the softening of its matrix.

During World War II a shortage of strategic metals suggested plywood as an alternative. The British Mosquito fighter bomber and the American PT boat are two high-speed plywood successes, hot-molded over forms in massive pressure vessels. After the war these great autoclaves became available and produced boats on both sides of the Atlantic: in Britain the Atalanta class, in the United States the Thistle class and the Wolverine Wagemakers. Excellent boats, their production was not without problems because the molds were solid but expensive, because the pressures were high (50 to 75 pounds per square inch), and because the great autoclaves that applied those pressures with high heat were aging and economically irreplaceable.

The Gougeon brothers of Michigan developed a new plywood system that was more than a refinement of wartime successes. It was a new way to think of wood, the W.E.S.T. system: Wood Epoxy Saturation Technique. This system uses wood's basic porous nature to absorb a thin liquid epoxy solution that hardens within the wood, excluding moisture, avoiding hygroscopic instability, and increasing its strength. This is a cold molding process of successive veneers laid and stapled over a light form (the staples are to hold the form temporarily), each veneer impregnated with 20 to 25 percent epoxy.

The fascinating thing about the W.E.S.T. system is its new look at an old material, a fresh and successful use for boats from fourteen-foot canoes to fishing vessels of several hundred deadweight tons. It builds strong boats . . . and what else can it build? And what new insights are waiting in the boles and rings of our old friends in the forest?

A Cedar Creek Canoe cold molded using the W.E.S.T. System: a light ribless hull of red cedar with ash thwarts and spruce gunwales and keel; simplicity and enduring grace

pine, and finally copper. Skill in designing ships with good ventilation around the timbers and good protection for end grain wherever possible, skill in building a tight ship, and skill at maintaining her keep old wood boats afloat: skill.

that is tight as a drum at the mooring, but leaks like a basket in a seaway. Well, before we jump to any conclusions, we need to remember that in all likelihood, the boat in question was as tight as a drum when she first touched water. There's no need for a properly built wooden boat to leak at all, and there are plenty of fine examples with dusty bilges to prove it. But for various reasons, some boats, large and small, begin to work themselves apart as the years go by. Not that they *must*, but that they *do*. The causes for such working can be anything from fastenings that are too light or are corroded, to broken frames, to poorly fitted floor timbers, to a design that failed. The root causes are usually simple and unnoticed by the average owner. Yet, as the hull begins to work, the effect is felt elsewhere in the boat as that thrust is transmitted throughout her. But faulty elements don't mean that the boat is a hopeless case, any more than two broken legs mean that a man will die.

What is not always clear, however, is that the price of repair — in the hospital or the boatyard — is usually unpredictable, usually high, and requires a skilled artisan whose temperament can make the difference in the job. But remember, we're talking about living beings here. We're talking about love and trust, confidence and courage. If you attempt to care for your wooden boat without these things, she becomes a burden to the spirit. If she is not understood, and loved for what she is, you'll never be happy with her. The parallels are surely obvious.

Most of us are willing to let life teach us much, and the lessons we learn aren't always so comfortable. Sometimes, although we see ourselves as salty romantics, born to the sea, we discover that life on the sea is just like life ashore, and find ourselves unfulfilled and sometimes frustrated by the burdens of care and responsibility for our boats. Perhaps we need a change.

Certain boats need certain owners. The ownership of a tired little vessel demands an incredible will. A will that takes over as the structure weakens, a resolve that can never fail. And it's perfectly reasonable. Some boats are held together by the will of their designers and builders. Those which are not must depend on their owners. They are never separate from their owners, but extensions of them.

The soul of the wooden boat is only intact when its elements are. Living close to the boat gets one deeply in touch with it, and care and attention yield beauty and pleasure. We, the owners, are integral parts of that soul, and no boat survives without love. But, neither does any other living thing.

Editors in chief are supposed to at least own a three-piece suit. Preferably they should have silver-gray hair, a musty education, a townhouse, and a detached air. Jon Wilson, then, cannot be the editor in chief of WoodenBoat: he is small, dark haired, gets excited about quite a lot, sits sideways in overstuffed chairs, and looks as if he could gnaw through a door if it were locked. He is obviously a front man for the real editor in chief. Whoever does the work is good: WoodenBoat is beautiful. Jon is such a kind and delightful man that, if he wants to tell me he's the head man, I'll humor him.

Wood is a living material. After it seasons, that is, after its individual cells die and their working moisture passes from them and the cell walls harden, timber still holds a pulse, a breathing in and out of the air's moisture. This respiration causes the wood to expand and contract. In some woods the change is measurable and troublesome and powerful. Roman quar-

Wood and Small Craft

by John Swain Carter

This subject is too large. The Navy Department has published an entire series of manuals concerning the topic. Many books and articles have questioned wood's viability as a boatbuilding material in our age of mass-produced plastic and wonder metals. This, then, is only an introduction to the subject, and those who want comparison will be obliged to look elsewhere, for in my mind there is no other suitable material for the purpose of constructing small craft.

This choice rests on many considerations and goes beyond structural properties. Wood is an organic material, a renewable resource, and this is an important consideration in a culture that uses finite resources for purposes of convenience. Wood has been tested through the centuries in all types of vessels and proved true. Its continued use provides us with a remarkable tether to our past, for the technology involved in wood production hasn't changed significantly through history. Thus, awareness becomes a part of boat building and is reinforced through the use of traditionally evolved designs. Viewed in this light the process as well as the artifacts are linking devices, within and across cultures, which can be useful in

our self-evaluation and determination of our future.

Wood is a simple, basic material to work. Its use isn't governed by any magical laws. The available properties are visible to the user and not buried in a series of charts and tables. Weight and strength are equated by the eye, likewise unhealthy conditions are plainly visible and often determined by an aberrance in natural color. Recognizing these properties in order to utilize them is primarily a matter of experience. Learning basic wood types and basic wood technology is a first step toward proficiency.

Most timbers suitable for boat building aren't available at the local lumberyard. This means a trip to the sawmill and an ability to cull through the available boards for the best construction (see the article on sawmill practice and wood use, on page 17).

The two properties that are most desirable are strength and light weight; durability and beauty will usually emerge through their judicious use and a craftsmanlike approach. The best and most easily available North American wood that provides this strength and durability is oak. The two varieties most commonly used

rymen drove pine dowels into drilled holes along their splitting line and poured water after them. The pine absorbed the water and swelled with enough force to break out great blocks of stone. Over time the expansion and contraction of breathing wood can work joints loose, expel dowels, loosen fittings, crack and craze finishes. Uneven absorption or late sea-

are red and white. White oak is considered better, as it is the most rot resistant, although both average about the same in strength and holding ability.

White oak, known variously as chestnut oak, swamp chestnut, and post oak, is available in most areas. The color of the sawn board varies from a yellow to a grayish brown, and it is often difficult to distinguish from red oak. Two small characteristics can help you discriminate: the pores in the freshly sawn end grain of white oak are extremely minute and usually filled with a substance collectively known as tyloses, while red oak has relatively large, unblocked pores, which account for the rapid spreading of rot; the color differences between the freshly cut woods can sometimes distinguish the two (though this is often unreliable). Red oak usually has a pinkish color throughout the grain and a heartwood of a deep brownish red, while a fresh-cut white oak board has the light coloring mentioned above and a heartwood of deep brown.

Oak is used as the strength of a boat: the keel, frames, stem, and transom. It is also used where durability is an important consideration, such as in the gunwales and plank-sheer.

Sharp tools are an important consideration. Oak is a tough, fibrous wood with mineral content that dulls edges quickly. Hand planing should only be attempted with a sharp plane scrubbed at about a 45-degree angle across the grain. Of course this is often impossible because the grain seems to run in several directions. Work slowly around the grain using a little bite with a jack or smoothing plane. Try

to select boards with grain suitable to the piece you are making; make patterns of cardboard or thin pine and try them on several pieces of stock.

Frames for small boats are usually steam bent. The best oak for the steam bending of frame timbers comes from the butts of young straight trees about 10 to 20 inches in diameter. The timbers should be sawn with the grain, clear of knots and other irregularities. Steam makes the tough oak frames supple enough to bend in place. A steambox is a woodbox or tube long and large enough to contain the timbers, into which steam is introduced by one means or another. Some advocate the use of exotic substances in the water used to make the steam and these are supposed to speed the process or make the frame more supple. Hogwash; steam is all that's needed to make the oak supple enough to be forced into place.

Framing or timbering out is usually a two-man operation. Make a number of extra frames to replace those that fail in bending. Opinions vary about green versus seasoned wood for frame stock and they likewise vary with different lots of oak. Personally, I have had better success with fairly green wood. Oak used in other areas of the boat is most often best if it is as dry as possible. Dry in this sense means a moisture content hovering around 15 percent (what you would expect of a seasoned piece in a dry shop). If a vessel is built with the seasoning of wood in mind, seams will have a better chance of staying tight and board ends should not split as easily, since the wood will have shrunk as much as possible for most con-

soning and other internal strain caused by dimensional change can warp the lines of wooden furniture, houses, or ships. Wood can crack — or *check* — across its end grain. Some lumber is unsuitable for any construction because of its twisted grain, which, as it breathes in and out, will twist any piece out of true

ditions. There are always exceptions; that is part of boatbuilding. I once helped cut out an oak log for a lobster boat keel in the morning and by afternoon it was having the molds set up on it for framing. This keel was eventually painted liberally with red lead primer and didn't check or warp, but the practice isn't recommended on smaller craft.

Another major component of the boat is in the skin, or planking. Several considerations of wood selection are important here. Weight figures predominantly. Strength is in the fit of one plank against another, and in the fit of planking against the frame and other integral parts. It is not necessary to use a wood with the strength (and consequent weight) of oak. Planking woods are chosen for lightness, availability of long lengths, shock resistance, and lack of structurally debilitating knots. Easily available woods that meet these criteria are the northern softwoods, primarily cedar and pine.

The cedars are the most useful if not most abundant of the planking woods available for small-boat construction. They fall into several common groups. Atlantic white cedar and northern white cedar are similar except that in northern we find more knots, requiring careful selection. Port Oxford and Alaskan cedar are the exotics. They are West Coast woods and both trees produce clear, long lengths ideal for planking — although cost, due to their limited supply, makes them the ideal rather than the rule.

Northern white pine has also been successfully used to plank boats. It is particularly use-ful in planking small rowboats and canoes because of its cheapness, workability, and the low ratio of knots to length.

Planking is an art that is acquired with time and patience. Some rules of thumb have developed over time. Practice has shown that the thickness of planking should vary in proportion to the length of the boat being planked. This is not a hard and fast rule but it is useful to keep the following proportions in mind:

10-foot boat:	¼-inch planking
20-foot boat:	½-inch planking
40-foot boat:	1-inch planking

Again, lightness and durability are considerations and plank thickness will vary depending on the intended use of the vessel. A boat that will see rough beach service, for example, will out of necessity have thicker planks than a Rangeley lake boat that will be portaged frequently.

Planking is also one of the most noticeable elements of the vessel. It reflects the builder's craftsmanship and pride. Care should be taken in planning, and a carved scale model is an excellent way to determine plank lines, drawing them out on the model's surface. Battens, long thin strips of wood, can also be tacked to the molds to determine the plank lines.

Finishing the boat after it is framed and planked also requires forethought. Many a good boat has been ruined by poor judgment when it came to installing the thwarts, floors, gunwales, and other finish features. Maynard Bray, technical editor for *WoodenBoat* maga-

with irresistible force. Skill in choosing the right wood for the job, skill in seasoning it, skill in working with its grain and planning for its slow but inexorable movement and working with that, too, and skill in finishing and protecting wood from rapid changes keep wooden things whole and true: skill.

"WOOD FLOATS"

Boats and wood are natural companions. Wood is the shapable, workable, changeable, resilient stuff shipwrights still need. There is an old understanding between wood and the water, an old contract that breaks down occasionally but holds up remarkably well for an agreement as old as the first tools. Cabinetmakers and house carpenters can learn chapter and verse from the shipwright, whose trade has a long research and development period beginning before Queen Hatshepsut's royal barge and not yet complete. Likewise, the shipwright can learn a few strokes from his dusty land brothers and sisters.

The literature of boat building is a huge body of information, well kept and up to date. No single book could appraise it, and in lieu of any partial appreciation I will mention only three or four sources of other sources.

The International Marine Publishing Company is involved in boats and boat building as no other publishing house. Their list of marine titles and books on woodworking would bulk to several pages and can be seen each month in the interesting pages of *National Fisherman*, a newsmagazine that most of both coasts look forward to. *National Fisherman* regularly runs articles on boat history, boat building, and some woodworking features.

Under their imprint is the *Mariner's Catalogue*, in several volumes, a dense and enjoyable catalogue of catalogues, comments, access and images that will fill the empty months between pulling your boat and launching her.

The most beautiful magazine of boats is surely *WoodenBoat*, published in Maine by a crew under captain Jon Wilson (see page 69) who fit and join each issue as carefully as Donald McKay's shipwrights put together a clipper, often as beautifully. Not really — nothing has been as beautiful as the clippers — but *WoodenBoat* deserves the compliment, and your subscription. Regular features on woodwork, woodworkers, wooden boats, tool reports, new plans, workshops, launchings and — in the case of a few masters like Pete Culler — departures. I love *WoodenBoat*.

David R. Getchell, editor in chief
National Fisherman
Diversified Communications, Camden, Me. 04843

Monthly tabloid
102 pp.
$1.00/issue, $10.00/year

David R. Getchell, editor
Mariner's Catalogue
Published jointly by National Fisherman and International Marine, Camden, Me. 04843

5 vols.
190 pp.
$4.95 to $7.95
Illustrated

Jonathan Wilson, editor
WoodenBoat
WoodenBoat, P.O. Box 78, Brooklyn, Me. 04616

Bimonthly
120 pp.
$12.00/year
Illustrated

Jay S. Hanna
Marine Carving Handbook
Camden, Me.: International Marine Publishing Co., 1975

$6.95
92 pp.
Illustrated

Quarterboards, trailboards, billet heads, and railending are covered, and there's a good treatment for decoration but hardly a thing about the hardest part, carving letters.

zine, has begun an excellent series of articles on the care and consideration of these forgotten details. The series should be useful, but experience is imperative. Look at other boats, study them and learn how they are put together. They are straightforward, for the most part, and offer their construction details to any who will take time to look. When you see a pleasing feature that works well remember it or — better yet — sketch it out and tack it over your workbench for constant reference. Don't walk by a well-designed and -constructed boat without learning something from it.

One final note on finishing concerns economy. People often think they need exotic woods from faraway places for finishing their boats. They don't realize the beauty and utility of woods indigenous to their area. Fruitwood trees, especially apple, have excellent crooks for small knees and breasthooks. Also, their

Yankee in me lays aside any more little pieces, I won't be able to get into my shop — that's a chance I relish taking.

Unfortunately many of the books written on boat building don't go into the topic of wood as a boat-building material very deeply. Some do offer snatches of information, though; for further information, see:

Robert Seward, *Boatbuilding Manual* (Camden, Me.: International Marine Publishing, 1970).
    A good book for the beginner, this volume covers much material. Seward discusses wood types and uses briefly at the beginning of the book.

Peter Cook, *Boatbuilding Methods* (New York: Granada, 1977).
    This is a British book that offers many tips. A short section on buying and selecting timber is included, but its value for American readers is limited because of a difference between English and U.S. buying and selling practices.

Walter J. Simmons, *Lapstrake Boatbuilding* (Camden, Me.: International Marine Publishing, 1978).
    This book has a section entitled "Lumber and Tools." Simmons, an experienced builder, goes into practical aspects of the purchase of boat timber and gives good advice on the use of planking stock, natural crooks, and pattern stock.

Wood: A Manual for Its Use as a Shipbuilding Material, vols. I–IV (Washington, D.C.: Bureau of Ships, Navy Dept.).
    Out of print but available at most university libraries that are government depositories.

wood in suitable lengths finishes beautifully for a number of other uses. Likewise butternut, a favorite of the late Pete Culler, is a warm, mellow, shock-resistant wood useful for items like rowlock pads or thwarts. This is only the beginning of a list of other local woods that should be investigated and utilized. Look in your firewood pile, get to know these woods, and plan ahead, laying a little piece by the side for later use. My wife claims that if the

No tree is like another; within the kingdom of trees, within families of trees, within
a single copse of the same species of trees, not
one is like another. Growth through weather
and season has made each unique. Wood is not
a homogeneous material; each plank has its
own grain, its own knots and checks and voids,
each is a separate problem. Every species of

Oars

by John Swain Carter

The techniques for producing good oars for
small craft have changed little over time. The
best are still handcrafted of wood. With a
smart, well-designed boat, they provide a
cheap and wholly satisfactory way to healthful
exercise, bringing the user in close contact
with the natural world.

Many don't realize the practical pleasure
that can be derived from rowing. They are the
victims of poorly designed, mass-produced
rowboats and the short, heavy oars that dom-
inate today's market. These oars aren't limber,
their blades seem as thick as they are wide,
and their grips may have been made for King
Kong. Industry, which designs for cheap mass
production, has taken over the planning for
many products, creating uniform articles that
have lost the subtleties of the original object.

Fortunately there are still oases of craftsmanship in the production of oars. Shaw and
Tenney in Orono, Maine is one such refuge.
The firm has been producing well-designed
handcrafted ash and spruce oars since 1858
and I've used their beautiful work for some
time. They offer high quality, satisfaction, and
an efficient means of cheap propulsion.

One component of quality oarmaking is a
craft tradition that resists high mechanization
in favor of an individually created product.
Shaw and Tenney has adopted labor-saving
machines in the basic operations of cutting out
the oar blank, but shaping and balance are
determined by eye and each oar is finished by
hand.

The stock used is New England white ash
and northern spruce. Patterns developed and
passed down through years of experience are
drawn out on the wood stock with consideration for wood grain and the elimination of
weakening knots. The pattern drawn is then
cut from the stock on a bandsaw, producing
an oar-shaped blank. These blanks are put on
one of Shaw and Tenney's ancient oar lathes,
which fashion the grip and tapered loom of the
oar. The blades are then rough-shaped on a
large table saw and the whole oar is finish-
sanded by hand on a revolving sanding drum.
The finished oars are sealed with lacquer and
varnished (except for the grip, which is left
unfinished). The company offers leathers, buttons, and copper tips either attached or unat-
tached at extra prices.

Good oars, like all handcrafted products, are
expensive. There is a way to circumvent this

tree has its own virtues and its vices, a particular aptitude, a special strength, applications of marginal value, a range of poor uses, and many prohibitions. Willow accepts brutal shocks and makes a fine cricket bat or polo ball; a willow fence post couldn't last a season. White cedar makes long-lasting fence posts, light and durable canoe frames, and weather-

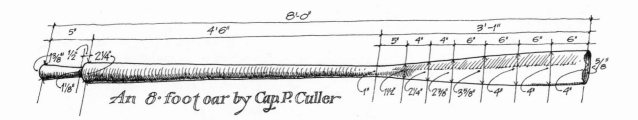

An 8-foot oar by Cap. P. Culler

problem: by making your own oars and learning something in the process. Oarmaking isn't that difficult and plans can be bought, copied from another oar, or made up. Making your own will help you to become familiar with tools and materials as well as allow you to produce an efficient, practical product from the natural surroundings.

Materials for oarmaking are more available than most people think. Straight-grained 2-inch spruce staging planks can be located at any lumberyard with some culling of the stacks. Ash can be found at most sawmills or mills that produce wooden pallets for industry. Length and thickness are not that important, as strong oars can be glued up using epoxy glue. Some important considerations are wood grain, knots, balance, and moisture content.

Wood grain should be carefully examined so that the finished oars don't break or split in use. Common sense is the best adviser in this area because wood is basically a straightforward material.

Look at an oar or picture it in your mind. It is essentially a lever and the grain should run straight down the loom and blade. Oars that are glued up from thin pieces gain strength by opposing the grain of the laminant. When it is time to shape the looms using an edged tool, however, opposing grains can be hard on the nerves, especially when it comes to planing. Large knots that will weaken the oar should be avoided but smaller pin knots are not particularly bothersome in this respect, but their contrary grain will be difficult.

Dry wood — that is, wood with a moisture content of 15 percent or below — is a must in oar production. If dry stock has been left outside or shows surface moisture, bring it into the shop and allow it to dry thoroughly. Dry wood is essential for two reasons: green or unseasoned stock will cause the oar to warp as it dries because wood cells change shape as they lose water. If you are planning on gluing up stock to make your oars, the wood must also be dry.

A band saw or jigsaw can cut out the pattern-marked stock. A draw knife and spokeshave, with or without curved blades, will help rough out the oar and begin the finishing process. A spar plane will also be useful. As the oar begins to emerge from the stock, keep several things in mind. One is that you are making a pair of oars that should be as nearly bal-

proof shingles, but it splits too easily to stand shock or be of structural value. Skill is essential in seeing the strengths and faults of each individual piece of lumber, in selecting a wood for the purpose, in using wood at its best: skill.

90

anced as possible; work on one, then the other, always comparing weight and shape. If you are making your first pair, have a finished oar that you have used and like at hand to study and make comparisons. Balance in the individual oar is also important. One thing that a tapered loom provides is a concentration of weight toward the grip of the oar. This is desirable and some oar patterns (such as the Adirondack and the Culler oar) concentrate mass at the grip end of the loom. This puts the heft toward the inboard part of the oar and helps to reduce fatigue in lifting and moving the oars. Added weights and a difference in balance between oars may seem minute, yet these two factors can compound overtime when rowing, to cause discomfort and weariness.

Balance is a delicate phenomenon; don't carry inboard weight of the loom or any other feature to extremes. Always think of application in designing and making a set of oars. Form and function should be intermingled; a lightly built whitehall deserves light, springy oars just as a loaded dory needs long and strong oars.

Finishing techniques vary. I prefer a varnished oar that is sealed just with a commercially available sealer. The grip should be left sanded and natural, as this is more comfortable to the hand. Linseed oil and paint are other possibilities. There are various ways to protect blade tips from splitting. I mentioned that Shaw and Tenney sells copper blade tips that can be tacked on. Epoxy resin can be used in conjunction with fiberglass cloth over the end of the blade. Leathers and buttons are also wise protection, because shifting wood against a metal oarlock can quickly break down the wood fibers.

Handmade, well-balanced oars should be the goal and this concept should be with you from design to finish. Constantly examine and try other oars, study them carefully, and note how they are made. This is essential, for we too often look to books like this to enlighten us about craftsmanship. Books offer advice and help to cajole us into starting a project but this kind of learning is best tempered with experience. The feel of a wood grip in your hand, leather sighing in the oarlock as you feather and then pull away with legs and back will tell you more about what an oar is about than a whole encyclopedia on oars.

A steady hand and knowing eye become coordinated with useful tools and materials to produce an oar only when we decide to take the time and make an effort toward that end.

Further information on oarmaking:

Mystic Seaport offers a printed instruction sheet and a one sheet plan for a style of oar designed by Pete Culler. Contact

Oaratorial Dept.
Mystic Seaport
Mystic, Conn. 06355

Walter J. Simmons, *Lapstrake Boatbuilding* (Camden, Me.: International Marine Publishing, 1978).
This excellent little book contains a short section on oars.

R. D. Culler, *Boats, Oars, and Rowing* (Camden, Me.: International Marine Publishing, 1978).

Pete Culler packed a lot of information into this short book. He discusses his philosophy and tool preference in a chapter on oars and oarmaking.

Places to obtain quality manufactured oars and paddles:

Shaw and Tenney
20 Water Street
Orono, Maine 04473

Price list on request; 5- to 12-foot oars in ash and spruce;

4½- to 5½-foot paddles in spruce and ash

Swanson Boat Oar Co., Inc.
Albion, Pa. 16401

Price list on request; 4½- to 16-foot oars in white ash, poplar, basswood, and maple; 2- to 12-foot paddles in white ash, poplar, and basswood; boat poles

Dafron Industries, Ltd.
Blockhouse, Nova Scotia

Price list available on request; oars and paddles

Between the vices of wood and its virtues is the gray area of care, the province of craftsmanship in which the medium is used with wisdom and caution as a rich resource with special difficulties. This gray area spans all human endeavor — mountain climbing, painting, architecture, writing — a level of thought and action between enjoyment and difficulty, a resolving of problems toward an end. Skill

Ojiisan: A Traditional Japanese Joiner

by Carol A. B. Link

In Japan, joiners are placed in separate categories depending on their speciality. The *tansuya-san* specializes in making large *tansu*, wardrobes used to hold clothing. The smaller and more varied pieces are made by a different specialist, the *sashimono-shi*. Although both *tansuya-san* and *sashimono-shi* are joiners, the *sashimono-shi* is regarded with deference by connoisseurs in general and by the *tansuya-san* in particular. This is because *sashimono-shi* exercise their skills over an enormous range of products, have a very high degree of versatility, and display great mastery of their craft.

Sashimono-shi are now rare in Japan, but one of the best, Mr. Yusaku Tsuzuki, still lives in Kasukabe. He is a *bijutsu-sashimono-shi*. The whole term can be translated as a joiner whose level of skill is so perfected that the cabinets produced are considered to be works of fine art like paintings or sculpture. At the present time, there are only four or five *bijutsu-sashimono-shi* and they are all older men ranging in age from sixty-eight to ninety-six. We addressed Mr. Tsuzuki as *Ojiisan*, a respectful term for an elder man or a grandfather.

Mr. Tsuzuki's son, Yukio, is his only apprentice. Yukio, as a *mokkogei deshi*, makes *mokkogei sashimono*. The first term can be translated as very skillful artisan joiner. It indicates his status and means he has not yet mastered the craft as well as his father. Of course, this does not mean he is inept, as he is still considered to be much higher in rank than a common joiner.

A five-minute walk from the railroad station will bring you to the home and workshop of the Tsuzukis.

When Ojiisan, who is now seventy-four, was about twelve years old, he finished primary school (six years in Japan) and was apprenticed to a *sashimono-shi* in the old downtown section of Tokyo. He finished his apprenticeship when he was about twenty-one years old. He then returned to Kasukabe, set up his shop, and married. He has lived in exactly the same place for over fifty years and followed in the same occupation for the past sixty-two years. His work is his life.

The focal point for Yusaku Tsuzuki and his family is the shop (*shigotoba*). Supporting a family of eight is a difficult job anywhere in the world, especially when one family member

is intelligent compromise between debits and assets and its real force is in appraisal. A machine blindly picks up a billet of unknown material with unfeeling claws and works it in one way, disregarding all irregularity: this is not skill. A sailor feels the wind change on his face and hears the slightly altered pitch of the wind curling around his ear, his eyes read the waves and his hand on the tiller steers the boat to

is a college student. Fortunately, the Tsuzukis never lack for orders and they are always busy. Unfortunately, they are not really well paid for their labors, though the prices for their products may seem very high to a purchaser. The simple fact of the matter is that Ojiisan and Yukio work almost continuously in order to support themselves and their family.

The shop is also the central arena for most of the family's social life. When the children chat with their grandfather during the day, it is in the shop; this is the place where wholesalers discuss orders and prices, itinerant peddlers attempt to hustle, family friends and relations are entertained, missionaries strive to proselytize, and news is exchanged by and about one and all.

The shop is an 8-*jo* (12-by-12-foot) room. It faces north and has sliding glass doors at the front and west sides. The northern face also

Ojiisan

use the wind to its limit; this is skill. A worker hefts a sheet of steel into a press, pins hold it at the proper position, the worker trips a switch and the sheet is stamped into a perfect form; where is the appraisal, the compromise of qualities?

Properly, it begins with a need. A man running from a boar has an immediate need: he lays hold of the first pointed branch he sees

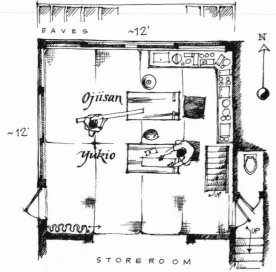

the Tsuzuki Workshop at Kasukabe
© Saitama Prefecture, Japan

behind the workbenches, there is a rack where *keshiki* (marking gauges) and planes are put away. Saws hang from nails beneath the stairs, where some small chests of drawers are also kept. These hold small tools, drawings of part of the Tsuzukis' repertoire of products, account books, pencils, small items that might be handy some day, and so forth. A box hanging from a pillar near the chests of drawers holds drills, a short ruler, and a small saw.

There are low incandescent lights in the room, hanging on drop cords from the ceiling. Both Ojiisan's and Yukio's lamps cast light over their respective workbenches from their right, but Yukio's lamp has a shade that prevents light from being cast toward Ojiisan's workbench. This is necessary because guidelines for cutting or planing are graved into the wood. These guidelines are usually very shallow and the worker is guided not so much by

has very low, overhanging eaves. The south side has one door leading to a staircase and the rest is completely open except in winter when it is shut off by a curtain. An opening leads into the next room where materials, tools, and so forth are stored. On the east wall, there is a ladderlike staircase going up to the second floor, which is also a store room. The floor, which is about one foot above ground level, is covered with straw mats. The ceiling is quite low, only a little more than six feet high. Natural light is constant and the room is very well ventilated.

There are some shelves near the ceiling on which glue, gauges, pegs, and assorted small items are stored. Along the east wall, right

In the shop of Ojiisan

and turns to use it. This is use more than skill. He must have leisure to appraise the problem and resolve it in a design. Safe in a tree or between hunts he can rough out a functional tool for pigsticking. Without pressing necessity he can bend his perception and dexterity to purpose, choose the strongest species for his uses, the straightest sapling; he can shape and fire-harden a point, compromising between tip

the lines themselves as by the shadows they cast. Thus, if the light is too bright or is shadow free, the work would be difficult, if not impossible. Accordingly, the lamps are shaded so that light comes from one direction only and workbenches are always perpendicular to a light source.

An apprentice would pay a fee to learn the craft from Ojiisan and would also have to buy his own tools while learning, a considerable expense these days, complicated by the fact that some tools are no longer made. He might

also have to do without dates and entertainment and provide his own insurance. The average young man looks at this situation and decides to study for a "good" school and begin the ride to success in a large company.

But Ojiisan has much to offer. One benefit is sixty-two years of accumulated skill and knowledge. Another is a home that is saturated with mutual respect and affection. He also has a daughter-in-law who is a good cook.

Other major benefits are work satisfaction and ultimate justification for one's own life. As

hardness and brittleness. His skill involves his knowledge of the sapling, dexterity with his tools, familiarity with the fire-hardening process, and the inner facility of suiting them all to his purpose, a utilitarian object, a good pig-sticker. This worker is close to his needs, he rides his life bareback, in contact with the movement and forces in it. Alone or in a small family his craft might progress no further than

Ojiisan and Yukio say, "You have a really good feeling when you can and do make things yourself," and, "There is nothing like the satisfaction you get when everything fits perfectly and is beautiful." I believe it. Their work also justifies their life because they observe a completed product and say, "I made this. All of it. It is part of me and will exist after I am gone. The very fact that I can and do make these things makes me a valuable human being." One only needs to compare this to the

aura of frustration and futility that permeates a large factory to assess the value of this benefit.

For the Tsuzukis the days repeat a pleasing pattern, which starts with breakfast prepared by Yoshiko, Ojiisan's daughter-in-law. After the meal, the workbenches, which had been put away the night before, are laid out and the day's tasks begin. Work is carried on, not in silence, but quietly. Conversation is usually about personal matters, news events and the

like. It may strike one as unusual that Ojiisan and Yukio do not normally talk about their work. This is because they each know it and know very well what to do next. If they do talk about the work, generally it is to decide which commissioned product to begin next rather than which step to do within the total process. Obaasan, Ojiisan's wife, washes the breakfast dishes and retires to her room to enjoy her hobbies. Yoshiko leaves for her job about nine and the children have already left for their day's tasks.

At about ten o'clock, Obaasan brings in the tea and *senbei* (rice crackers). This break serves two key purposes. First it is a relaxing, refreshing interlude in the work and one gets hungry after planing boards continuously for a couple of hours. Second, it is the first social interaction of the day. After fifteen or twenty minutes of tea and conversation, Obaasan leaves, washes up the cups, and busies herself about the house, while Ojiisan and Yukio resume their labors.

A little while later, Obaasan comes back to the shop and cleans up the wood shavings. Shavings can accumulate quickly and interfere with the work if they begin to bury the workbench, which sits on the floor. One of Obaasan's tasks is to burn these shavings and other scraps, so she often comes in, sweeps them up, and goes out to the garden to burn them. (This became the author's first and, for a long time it seemed, only task. The first stages in an apprenticeship involve cleaning and other "dirty work.")

this utilitarian level, a tool equal to the need. A larger society, a tribe, could give him more leisure, time away from imperatives. Without pressure he can afford to address needs below the surface and deal with them on another level of skill. Part of his need is now to communicate the feeling he has found in his leisure thought. He needs art. He might embellish his pigsticker with designs to make it

At eleven-thirty the radio is turned on to hear the news and the weather forecast. Work continues unless something of great importance or humor is being broadcast. If one listens very carefully, one can hear Obaasan setting the table for lunch.

Shortly after noon, Yoshiko comes racing home, greets everybody with her usual charming smile, and dashes into the kitchen. Presently, one smells pleasant fragrances accompanied by the tattoo of a knife on the chopping board. Yoshiko calls out that lunch is ready, and, after Ojiisan rises to go wash up and eat, Yukio follows. Lunchtime conversation usually has to do with family affairs: who has to do what at which bank; interesting news items; and other matters of daily life. After lunch, Ojiisan goes back to his workbench and reads the newspaper. Yukio may work on his *bonsai* in the garden or clean the fish tanks. Obaasan does the dishes and Yoshiko relaxes before returning to work by one o'clock.

At about three o'clock, there is an afternoon tea break, featuring some sweet potatoes that Obaasan has boiled on the old woodburning boiler in the garden. After the break, Ojiisan and Yukio resume work. Obaasan begins to prepare the daily bath for the family.

At about four-thirty, in the early twilight, Obaasan comes into the workshop, gets her pillow, and sits down. She stays there until dinnertime, maybe chatting, but often in silence, simply watching her husband and son at work. It is impossible to express adequately what is occurring at this time. Perhaps it is

sufficient to say that the silent communication is so personal and intimate that being a part of it is a great privilege.

Shortly after five o'clock Yoshiko returns from work and also stops to watch for a while in the workshop. She anticipates the future when she will take Obaasan's place here in the late afternoon, to sit, watch, and silently communicate with the others, while *her* daughter-in-law is fixing dinner. Envisioning this time, she leaves to do the daily shopping and begin cooking the evening meal.

While dinner is being prepared, Ojiisan and Yukio clean up the shop for the night. This involves putting away the tools, neatly arranging the materials and pieces of work along the wall, moving the workbenches to the next room, putting the pillows aside, and sweeping the floors. If they have time, Ojiisan and Yukio bathe before dinner.

Dinner is a pleasant affair, with as many members of the family as possible gathered around the table, and the adults drinking a glass of *sake*. After dinner Ojiisan and Obaasan return to their room to watch television. Yukio writes poetry and Yoshiko cleans, reads, and prepares everybody's box lunch for the next day. Everybody retires early to rest for tomorrow.

My life is not one of leisure; it is a life of occupation. Every day is like the day that went before. My workbench lies before me. My heart is like a mirror — all my work is reflected in it. (Yoshida Kenko)

uniquely his, an expression of self. He might charge his tool with the symbols of strength and cunning, transferring inner hopes to physical potential. He might shape improvements, alter the original form to suit better its function or simply to suit his (leisure spawned) idea of beauty. None of this embellishment or

Daily life is not a continuous round of drudgery. Each day is not exactly like the day that went before. It brings new problems, new work, different visitors, new and sometimes challenging commissions, different errands.

querers, fan makers and so forth. In their opinion, those engaged in what Westerners might call the "trades," such as carpenters, mechanics, and day-laborers, are not "real" shokunin.

Ojiisan's and Yukio's identity as shokunin is

The attitudes and values of Ojiisan and Yukio are based upon their self-proclaimed identity as shokunin. The characters for shokunin literally mean working person. According to Ojiisan and Yukio, the term shokunin correctly applies only to artisans such as joiners, lac-

reflected in their attitude toward their work and the world in general. They take great pride in working with their hands in order to produce directly a whole product. They feel pride of accomplishment in both the physical manifestation of what they have done and in

change would affect the basic purpose of the pigsticker. He continues to compromise the demands of his need with the abilities of his medium. The craftsman, who has now used his work as a medium for expression, as an art, is still close to life and necessities.

the skill with which they have done it. This was best demonstrated when they had a commission to make very small chests (8 by 8 inches) that contained twelve drawers. This was the first and, in all likelihood, the last time they would ever make this particular product. First times to do things are usually tricky, so there was some anxiety over the results of the project. Moreover, the object was for a special show at Japan's leading department store, Mitsukoshi. This fact produced more tension.

By the Tsuzukis' standards, drawers must be perfectly flush with each other, must slide easily with no strain and they must *not*, under any circumstances, rattle. A rattling, loose drawer is anathema to Ojiisan and Yukio. The first drawers were slipped into the chest with some trepidation, but their ultimately perfect fit brought an absolutely beatific smile of satisfaction to Ojiisan's face.

Those unacquainted with this work do not really appreciate this feature of drawers. The average person presumes only that a drawer should slide and the degree of perfection of sliding elicits a "so what?" reaction. Connoisseurs, cabinetmakers, their *shokunin* colleagues, and dealers seek out this feature and have a regard for the care and dexterity that goes into perfect drawers that surpasses mere admiration. It is a sensitive awareness and esteem for the skill of the maker and the devotion that he has shown to his work to produce a superb product. This is based on the knowledge that one cut of the plane too many will ruin the job beyond repair.

Pride in a difficult job well done, and in the esteem of those who can really understand and appreciate it, is a significant feature of Ojiisan's and Yukio's attitude toward their labors. It also reinforces their identification as *shokunin*.

This pride and identification is also manifested in other ways. For instance, Ojiisan and Yukio are, understandably, severe critics of both their own work and that of others. Ojiisan occasionally stops working for a few minutes and observes Yukio, just to verify to himself that Yukio is doing things properly. On occasion he advises Yukio about something or comments on his efforts.

Like the repeating days that make workshop life, the tools of the workshop have patterns.

Dai: benches

A *dai* is the main workbench. It is made of a solid section of zelkova, 36 inches long, 18 inches wide, and 4 inches thick. On the underside there is a brace that raises the front end about 1½ inches from the floor. On the top surface at the end nearer to the worker there are two bench stops at either side of the bench. Each is about 12.5 centimeters long and the center of the bench is open. This bench is used for almost all tasks and also supports the miter benches.

Craft is skill answering needs, focused by a sense of beauty. The needs are simple: food, shelter, comfort, protection. The answers to those simple needs can achieve an elegant beauty above cosmetics. It is a memorable experience to see a spoon or bow or chair that has the clean look of function and the look of the medium itself, the wood (or metal) speaking through the shape. Artifacts, single pieces

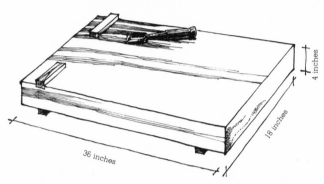

The *kezuri dai*, a planing bench, is a board of zelkova, 43 inches long, 6 inches wide, and 1½ inches thick. In use it is placed upon the workbench to the worker's left with the butt end braced against the left bench stop. It is used to plane long (greater than 36-inch) boards.

The *koguchi dai* (miter bench) is a jig like a bench hook used to plane edges of boards to a 90-degree flat surface. It is about 24 inches long and 12 inches wide. In use it is placed on top of the workbench against the bench stops.

The *tomedai* are jigs designed to produce 45-degree angles on either the butt end or the edges of a board. In use they are placed upon the workbench against the bench stops.

The *sankaku dai* is a mitering and jigging device that is about 10 inches long and has a triangular groove in it. Its purpose is to make triangular corner braces (glue blocks) for the stand of a cabinet. In use a one-inch square strip of wood, as long as or a little longer than the *dai*, is planed smooth on two adjacent surfaces. The smooth-surfaced faces are placed in the *dai* and the protruding material is

planed off. A triangular strip is the result. After this, the apex of the triangle is planed once or twice. When set into the corner of a stand, the lack of material on the apex allows excess glue to escape and thus forms a tighter joint.

Kanna: planes

The Japanese plane, or *kanna*, is a wedge-shaped blade inserted into a block of oak. Strictly speaking, *kanna* refers *only* to the blade; the wood plane block is a *dai* (stand). Planes come in three sizes with various-shaped blocks.

Mameganna (literally "bean blade") is very small, the blade usually less than 1 inch wide. The block ranges from 1 to 3 inches in length. These planes can be used in very small places, but the Tsuzukis rarely use them.

Koganna has a smaller blade ranging from 1 to 2 inches in width. The block ranges from 4 to 7 inches in length. These plans are used for the final finishing planing of the arrises of trays.

The *hiraganna* or (more commonly) *kanna* is usually about 3 inches wide. The block is about 12 inches long. This is the most often used plane. It is utilized to plane surfaces, to round off the arrises of a finished cabinet, and in the final finishing planing of most cabinets.

The *naga dai* has the same size blade as the *hiraganna*. The difference between the two lies in the length of the block. A *naga dai* is from 18 to 24 inches long. It is used to plane

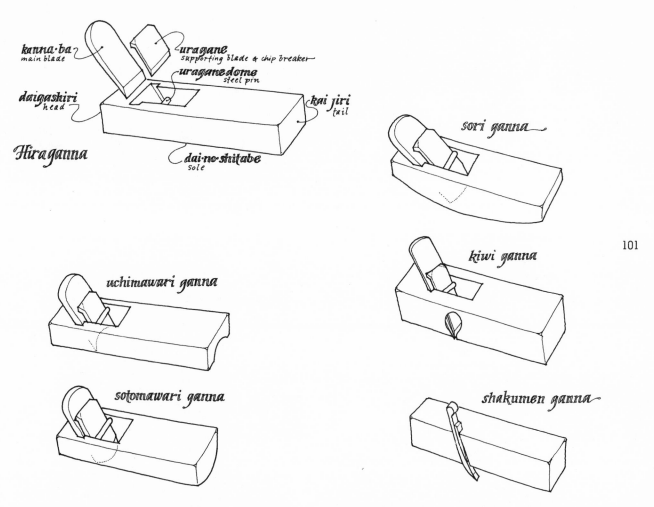

kanna·ba / main blade
uragane / supporting blade & chip breaker
uraganedome / steel pin
daigashiri / head
kai jiri / tail
dai·no·shitabe / sole
Hiraganna

sori ganna

uchimawari ganna

kiwi ganna

sotomawari ganna

shakumen ganna

the sides of boards and is equivalent to a fore-plane.

The cutting edge of the *uchimawari ganna* blade is concave. This plane is usually a *koganna*. Theoretically it should be used to round off the arrises of a cabinet, but the Tsuzukis seldom use it, as they claim that the controlled usage of the *hiraganna* gives better, cleaner results.

The cutting edge of the *sotomawari ganna* is convex. As with the *uchimawari ganna*, it is usually a *koganna*. The planes are used to plane the shallow groove that a door swings in.

The slightly convex cutting edge of the *sori ganna* (literally, warped plane) is in the size range of both the *mameganna* and *koganna*. The term *sori ganna* is more aptly applied to the block, which has a belly shape like a full-blown spinnaker sail. This plane is used to round the inside corners of door frames.

The *kiwi ganna* or "ledge" plane has its

blade set at an angle and fit into the side of the block. The blade ranges from one half inch to one inch wide. It is a rabbet plane and used to plane the edge of a board preparatory to making lap joints.

Gari dai is the Tsuzukis' personal term for this plane. *Gari-gari* is an onomatopoeic term for the scratchy or crunchy sounds produced while planing a deep groove. The closest equivalent in the West is the sound a car makes while going over a rough, rocky road. An alternative, and perhaps more standard, name is *shakumen ganna*, a spoonlike blade. This blade is very narrow, ranging in size from one quarter to half an inch. It is not actually spoon shaped but flat, and fits into the side of the block. It is a dado plane used to rout 90-degree dados in a board preparatory to making dado joints for inserting shelves.

The *uchimawari, sotomawari, sori, kiwi,* and *shakumen kanna* come in both the *mame* (bean) and *ko* (small) sizes.

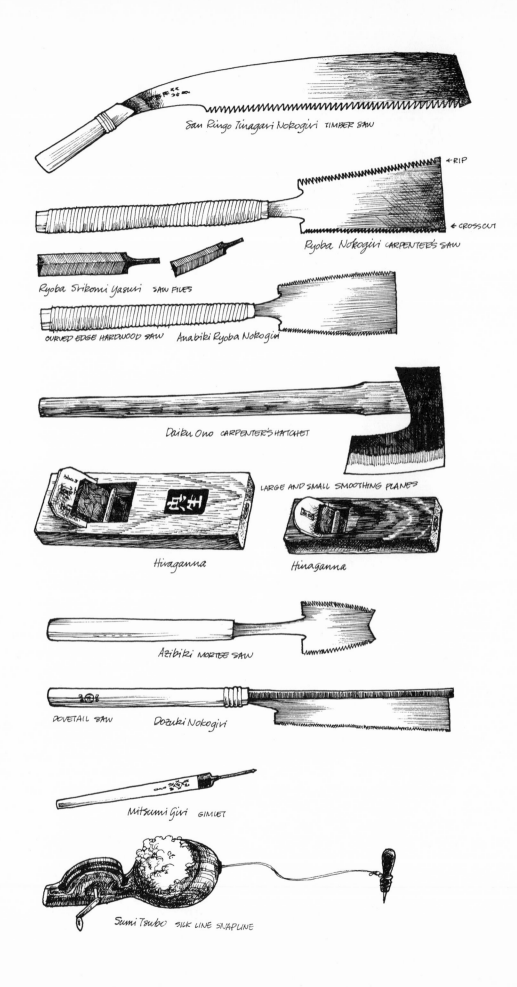

San Ringo Tunagari Nokogiri TIMBER SAW

←RIP

←CROSSCUT

Ryoba Nokogiri CARPENTER'S SAW

Ryoba Sribomi Yasuri SAW FILES

102

CURVED EDGE HARDWOOD SAW Anabiki Ryoba Nokogiri

Daiku Ono CARPENTER'S HATCHET

LARGE AND SMALL SMOOTHING PLANES

Hiraganna

Hinaganna

Azibiki MORTISE SAW

DOVETAIL SAW Dozuki Nokogiri

Mitsumi Giri GIMLET

Sumi Tsubo SILK LINE SNAPLINE

of craft old and new, have touched my mind, shifted my whole aesthetic, and have left images that will affect the way I see. What have you seen that is so pure and right for its purpose, so harmonious in its proportions and so apt in suiting its medium to its function that it remains in your mind as a yardstick against which other claims to beauty are measured?

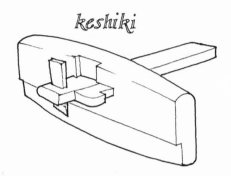

Nokogiri: saws

The shape of standard Japanese saws is significantly different from those of the West. In use, pressure is applied on the pull stroke. Saws vary in length, number of teeth, and length of handle. Also, they can be customized by a saw sharpener and special orders can still be placed for unusual saws.

The *tategiri*, or ripsaw, is used to cut along the grain of the wood. Ojiisan and Yukio seldom use this saw and it hangs from its peg on the wall, rusting away. Usually they can and do use a *keshiki* to cut boards with the grain.

The *keshiki* (gauge) is also known as the *kehiki* or *kebiki*. *Kebiki* appears to be the "correct" pronunciation in Toyko, but Ojiisan, Yukio, and other *shokunin* always call it a *keshiki*. Ojiisan, Yukio, and others have also said it is legitimate *shokuninben*, i.e., artisan dialect. The *keshiki* consists of three pieces of oak: a body, a sliding scale called a *sao*, and a wedge. A blade fits into a slot on one end of the sliding scale. These scales come in a great range of sizes. The largest is 36 inches long. After purchase, the user cuts the scale to whatever length is desirable.

The *keshiki* is essentially a marking gauge and it is used as such in the West, with a nail in place of the blade. In Japan it is used both as a marking gauge and to cut sections of a board along the grain. The Tsuzukis regularly use the *keshiki* in place of the ripsaw.

The crosscut saw, *yokogiri*, is used for all rough work where it is necessary to cut across the grain.

The *hozoshiki* is a rather finely toothed backsaw with a metal rib along the back to hold the blade stiff and straight. This saw is used to cut dovetail and finger joints.

Dozuki, a miter saw, is finely toothed with a metal reinforcement rib along the back. It is used for work where fine teeth and a lot of control are desirable.

The *kikugihiki* is a short (8-inch blade), coarse toothed, crosscut saw used exclusively to cut off the tops of the wood pegs that are used in joints.

The *itonoko*, or scroll saw, is supposed to be used for cutout designs in a plank. Ojiisan and Yukio can use this saw but they will not. They dislike the work it involves and they must stand up and bend over to use the saw properly. They say it gives them terrible back pain. Instead, they subcontract this sort of work to a jigsaw shop. They have done this for at least the past thirty-five years.

The keyhole saw, *mawashibiki*, is seldom used. Its main use is to enlarge the blade insertion slot in the *keshiki*.

The *hikigomi*, kerf saw, is a very thick bladed saw. It is custom-made and leaves a

James Krenov $13.50
A Cabinetmaker's Notebook   132 pp.
New York: Van Nostrand Reinhold, 1976 Illustrated

The Fine Art of Cabinetmaking $14.95
New York: Van Nostrand Reinhold, 1977 192 pp.
                                       Illustrated

To call James Krenov astounding would be to slight him: his work and his words are so subtle and human that hyperbole doesn't suit him. He is quiet and controlled and he seems to be more a thoughtful midwife for the ideas waiting in the wood than a master. If he can be faulted at all in these books, well illustrated with plain but illuminating photos, it is that he makes it all seem too simple . . . and he knows it. If you have any interest in wood you must have both of Krenov's books.

A good exercise in the difference between the skill involved in craft and the additional sensitivity that makes art would be to compare the powerful skills of De Cristoforo (*Complete Book of Power Tools*; see page 144) and the consummate artistry of Mr. Krenov.

104

very wide kerf. Its exclusive use is to cut a kerf for a slot in the corners of large trays. Later, a separate piece of wood is inserted in this slot for reinforcement purposes.

Nomi: chisels

Chisels, surprisingly enough, are not often used by the Tsuzukis. They use them mainly to trim off the protruding remnants of wooden pegs; to alter the throat of the plane block; and for mortise-and-tenon joints. However, mortise-and-tenon joints are only used in tables, and they made only two tables during the year I was there. It is not surprising that Ojiisan has been using the same chisel for about ten years.

Another very narrow (¼-inch) chisel is used to trim the ends of dado grooves.

Kiri: drills

The drill is a palm or twirl drill. It consists of a bit that is square in cross section and comes in various sizes. The bits and handles are purchased separately and one must insert the bit into the handle. This is a task that is much more difficult than it seems. The bit must be precisely in line with the axis of the handle or drilling will be a disastrous operation.

The *kiri* is used to drill holes for the insertion of wooden pegs in joints and at other points of stress in a product. The size of drill that is chosen is dependent upon the thickness of the wood being drilled.

Kogatana: knives

The knife, literally "small sword," consists of a blade with one bevel, about 3 inches long, set into a wooden handle and is used to carve out the inner corners of door frames. This is a job that requires care and judgment. Ojiisan always performs this task.

Natta: axe

The *natta*, a single-bevel shaping axe of 2 to 5 pounds, is used to trim bark and other blemishes from stock. These axes are not even sold in specialty shops and must be made to order, but most of the blacksmiths who made them are no longer alive.

Hammer and Mallet

The hammer and/or mallet are used for pounding in pegs, hammering nails and adjusting the blade of the plane.

Monosashi: rulers

Despite the fact that the metric system was officially instituted in Japan a century ago, Ojiisan and Yukio still use *monosashi* (rulers) that follow the ancient Japanese system of measures:

1 *rin* = .012 inch/.308 centimeter
10 *rin* = 1 *bu*

10 *bu* = 1 *sun*
10 *sun* = 1 *shaku* (.994 foot/30.3 centimeters)

Rulers come in one and three *shaku* lengths that have subdivisions marked in accordance with the above system. This system is firmly locked in the Tsuzukis' minds and all thought about measurement is carried on in terms of *rin*, *bu*, *sun* and *shaku*. The *shaku* is equal to one foot for practical purposes.

Nori, Bondo and Nikawa: glues

The Tsuzukis use three sorts of glues. *Nori*, which is made by grinding cooked rice to a smooth, grain-free paste; *bondo*, a commercial-grade white glue; and *nikawa*, an animal by-product glue. In the past only *nori* and *nikawa* were used but now they substitute *bondo* for *nori* for some joining tasks.

Nikawa is used almost exclusively for attaching the "hangers" inside *tansu* (wardrobes). It is strong and very fast setting. *Bondo* is used to glue individual pieces of stock together to make planks, and in laminating. Usually *nori* is used for all other gluing work such as in joints, lap joints, attaching stands, and so forth. One point to be noted is that *bondo* can be substituted for *nori* but neither can be substituted for *nikawa*.

Hata and Himo: clamps and cord

Hata (clamps) of one sort or another are essential in all woodworking to hold glued pieces together until they are dry. One set of clamps is made of iron and is used to hold boards that have been glued to form planks and to hold the finished cabinet to square after finger joints have been made.

The *himo* (cord) is used to hold pieces that have been glued together. They are wrapped with cord and the cord is tugged sharply at every angle of the piece. The cord is never cut and some leader is left between each item. Although it appears difficult and troublesome to bind pieces with this cord, it is actually easier to use than the clamps. The clamps are heavy and cumbersome. It also seems that you never bring enough with you from the storage room to the shop, so you must keep jumping up and down to get more. Also, you eventually run out of them when doing a lot of gluing. The cord almost never runs out, it is always handy by your side, it never falls off the glued piece, it holds the joints firmly, doesn't take much storage space, and doesn't rust. In total it is an excellent solution to the problem of clamping a glued joint.

Toishi: whetstones

Whetstones are one of the most important items in the tool kit. There are two of them, rough and finishing. The rough stone is a soft stone and essentially removes a layer of material from the surface of the blade to expose a new cutting edge. Use of the harder finishing stone hones this edge to razor sharpness. The edge that Ojiisan and Yukio put on their blades

Kip Mesirow and Ron Herman
The Care and Use of Japanese Woodworking Tools
Woburn, Mass.: Woodcraft Supply, 1978

$8.00
96 pp.
Illustrated

A more substantial book on Japanese tools will undoubtedly appear in response to the interest newly given the woodworking of Japan and China. Until then this slight and poorly illustrated book is the authority and is adequate, if disappointing.

Kiyosi Seike
The Art of Japanese Joinery
Tokyo: Weatherhill Kankosha, 1977

$8.95
126 pp.
Illustrated

This is more a philosophical work than a carpenter's manual but you ought to have it. Carpentry, Oriental or Occidental, is a philosophic pursuit of an active kind that needs strength, cunning, oversight, and inner direction; look at a good framer stopping with a cup of coffee to figure out the jack rafters for a dormer and see the face of a philosopher. Any good carpenter has it, but the spiritual direction of building is a formal part of the Japanese discipline, refined and polished over a thousand years. That's a lot of sawdust. The joints in these exquisite photos are physical flowerings of the Japanese thoughts about skill; taken with Kiyosi Seike's commentary they cannot fail to help your work. You may never cut a storksbill joint, but you will feel better about all your joints.

106

elicits fear and awe from Westerners. Usually accidental wounds are painless and bleeding will not occur until four or five minutes after the accident. The Tsuzukis' blades could easily be used for any sort of surgery.

Like all good artisans, Ojiisan and Yukio have a great deal of respect for their tools and equipment. But in Japan, for them and others, this respect goes one step further and is a religious ritual.

Shortly before New Year, Ojiisan and Yukio informed me that there would be many, many *kamisama* (gods) in their home and shop over the holidays. On December 31, we thoroughly cleaned both the first and second floors of the shop. We vacuumed, washed windows, dusted, and even swept the crawl space beneath the floor. Then all the tools, workbenches, the desk I used, drawings, pencils, etc. were neatly arranged along the east wall of the shop. On the evening of the thirty-first, Ojiisan expunged any evil spirits from the house, shop, and family by using the *gohei* (strips of specially folded white paper attached

to a piece of bamboo). Candles and offerings of *mochi* (rice cakes), oranges, and sake were placed on top of the arrayed tools. These offerings were left there for three days. The reasoning behind this is:

1. You cannot work without tools.
2. Therefore, the tools must be *kamisama*. (This is a very complicated conception of a god or spirit.)
3. So, we make offerings to these benevolent *kamisama* that enable us to live by their very existence in and as our tools and equipment.

Other *shokunin* performed this same ritual. It is somehow disconcerting to a foreigner to see these offerings arranged on top of drill presses, metal lathes and heavy equipment, but so they are. Fundamental cultural concepts can and do continue to be viable despite "modern" industrial trappings. This might be all the more true for *shokunin*, who may be more conservative and traditional than others. Modernization occurs in accordance with tradition and although some things may change, fundamental beliefs and values stay the same.

Dr. Carol A. B. Link is a student of technology and culture who is not satisfied with library work. She apprenticed herself to Mr. Yusaku Tsuzuki, who is called Ojiisan in this article edited from her large and careful doctoral thesis, "Japanese Cabinetmaking: A Dynamic System of Decisions and Interactions in a Technical Context." Her achievement in being absorbed into so distant and so rich a life and culture is heroic, though her manner is unassuming.

A Shaker rocking chair sits in Bert and Sylvia's kitchen. It is a quiet piece that does not shout for attention with carving or a fancy interplay of woods. As a matter of fact it did not catch my eye the first time I walked into the kitchen but over the several years I have been yarning in that room the rocker has insisted that I use it. It is pine, light and fine lined with a modest, companionable rock, not

The Workbench

by Jan Adkins

The way you work, the way you approach every step of a project, is most affected by a single tool, the workbench. At the beginning we can put behind us the notion that a workbench for woodcrafting is a flat surface with drawers under and pegboard over, as seen in *Better Homes and Gardens*. No, that is a potting bench. A workbench isn't a place, it's a tool. The woodworker has a basic need to hold work immovably at the most comfortable height and angle for the procedure at hand and a workbench is the tool to do it. Look at the benches and holding tools of other professions and see what a woodworker's bench should be. The gleaming sphere of an engraver's vise spins on its bearings and pivots on its leather, shot-filled bed to present work at almost any angle. The dentist's chair is a workbench: adjustable for angle, height, and lighting, with racks for tools and a swinging work table. A cobbler's bench holds cobbler, tools, and shoe in a given, comfortable relationship so he can retrieve and put back any tool without looking. The welder's workbench is a steel ground with racks for clamps and chipping hammer. The chef's and butcher's workbenches, overhung with tool and pot

racks, offer clear cutting surfaces that will not turn or dull the knives hung in racks at the sides. The woodworker's bench should be no less convenient or well thought out. Even the weekend handyman works at levels that need more than a countertop.

The professional cabinetmaker has always needed a proper bench, but the profession itself virtually disappeared in the United States for many years and many of its tools with it. Replacing the cabinet shop, its benches and handtools, the furniture factory machined components, using power to replace straining muscle, and its benches were platforms for assembly, like a mechanic's bench.

With the return of the woodworker as an interested amateur or, more rarely, a commissioned artist, the specialized tools and workbenches of the trade have reappeared in catalogues and in the shops and dens of the tool wonk.* In this reincarnation the workbench has acquired an unnatural dignity. This homely practical tool has become an altar, and the amateur woodworker approaches it

---

* A *wonk* is a vocal and ardent bore on any subject whose field of focus has narrowed to include only the purest and most distilled version of the pursuit.

a swooping camel-ride. I could pick it up with one hand and hold it at arm's length, yet it supports my nervous, fidgeting bulk solidly. It is almost a hundred years old. The only flaw I can detect is that the turned piece that caps the forward leg and support over the arm rest is loose. It turns on the spindle and speaks to me that way. The design is so plain that it has taken me years to realize how beautiful it is.

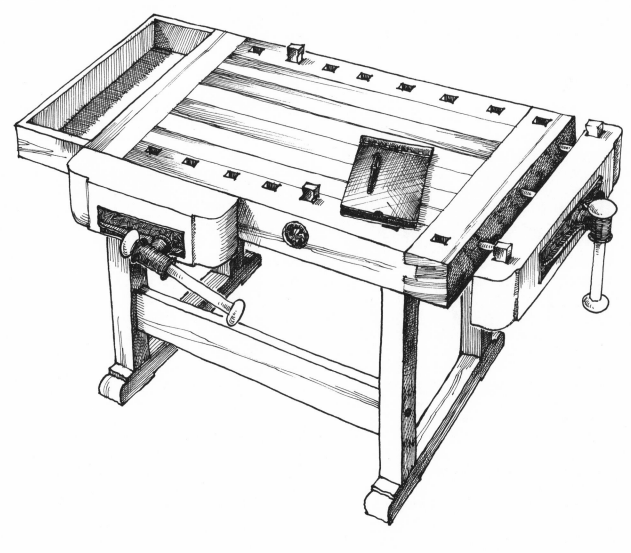

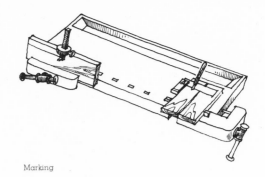

Marking

Black & Decker Workmate Workcenter

too reverently. He is more concerned with keeping his bench pristine than with putting out simple, useful work. One woodworking magazine recommended a square of carpet laid on the beech surface of a workbench, so those nasty tools wouldn't scratch it.

Many professionals work on benches no more imposing than a plywood or masonite surface on a heavy base, fitted with one or two good vises and cut to accept bench dogs (the wooden or plastic or zinc stops that fit into holes and hold work against the countering vise dog). Dean Torges (see pages 134–138) prefers a Homasote surface, believing that his unfinished pieces should not be burnished by harder materials. Some workers want an uncomplicated, replaceable surface into which they nail various jigs and holdfasts without pious guilt. They depend on a variety of clamps and wedges and machines like the Zyliss vise to hold their work. The Japanese carpenter (this is too humble a word, see pages 92–106) works at the simplest slab of zelkova fitted with dogs; he uses wedges, clamps, and cord to hold fast. In the field, a Japanese house carpenter (again, it should be *daiku*, a word identical in roots to our "architect") uses a post driven into the ground and a movable shoulder dog lashed to the post. They do very well.

The obvious point is that you can work well without a six-hundred-dollar beechwood Ulmia bench. Krenov uses such a bench and treats it with great respect. That respect is part of his art and he is so assured that it will not hamper him. For a *sashimono-shi* it would be a disgrace to *step over* any of his tools, show-

ing rude disrespect to objects steeped in honor; in this fine joiner such elaborate respect is workable and seemly but I am not sure I could work comfortably with too many rules of tool etiquette. I lay my planes on their sides and if I am disrespectful enough to bollix up an edge I feel properly disgraced, but I leap over my best hammer without a twinge. A big pipe organ of a bench is just too great an investment not to intimidate the amateur and cow the wonk.

The subtler point, and the more important lesson, is that all good craftsmen — in their own ways and at their own levels — have devised some stable way of presenting work to their tools at the right angle and height. Can a workbench help you? Undoubtedly. Can you use a famous bench? That's up to your ego and

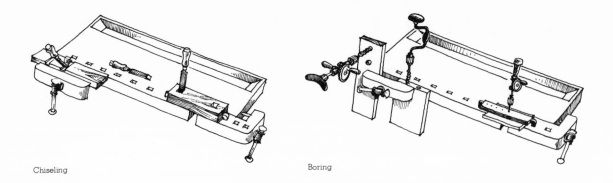

Chiseling   Boring

your checkbook. If you can approach it without incense and chanting, a well-made workbench is marvelously simple and convenient and will shorten the setup time for a lot of operations. You might find a bench that has already been sufficiently disgraced. Can you substitute for a big beechwood woodworker's bench? Yes, of course: the most informative project a learner could take on is to design and build his own bench. An intelligently built base of dimensioned lumber, triangulated and well-fastened, can be as stable a support as furniture-grade hardwood legs, and if the vises are of good quality and correctly placed to oppose snug dogs your work will not notice that the surface isn't rock maple.

The Black & Decker Workmate is not as large a surface as a full-sized bench, but it is an ingenious and useful response to woodworking needs. Its narrow worktop is split into halves that oppose one another on pivoted screw vises so that almost any shape (to a reasonable width) can be clamped between nonmarring plastic dogs or between the halves themselves. The entire unit folds to and stores behind a door. It cannot clamp long pieces, and tool space (solved in many benches with a level recessed below work height) is limited, but for occasional use, carpentry on site, or for excellent service with an accompanying table of the same height, the Workmate is a simple, inexpensive alternative.

Placement of a bench in your shop can affect its utility. A bench should stand free, with circulation all around it, to accommodate large sizes and unusual shapes and to give working access to every edge. A bench should not be placed against a wall like a counter.

Consider the uses of the workbench and the time you will spend hovering around it cutting,

Clamps and holdfasts

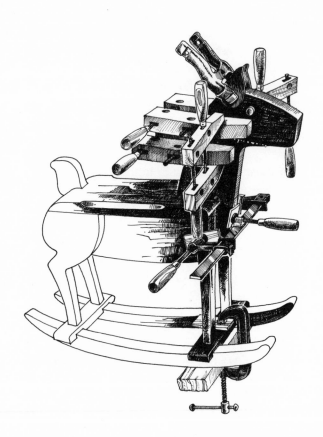

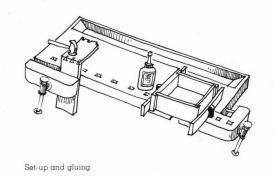

Set-up and gluing

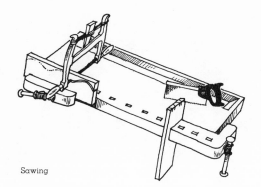

Sawing

planing, shaping, scraping, boring, sanding, assembling, and gluing. Consider the fatigue factor of working at an unsettled piece wobbling even slightly. Consider a workbench as the foundation of your labors and as a companion piece for all the rest of your tools, and then build or buy a bench worthy of your efforts.

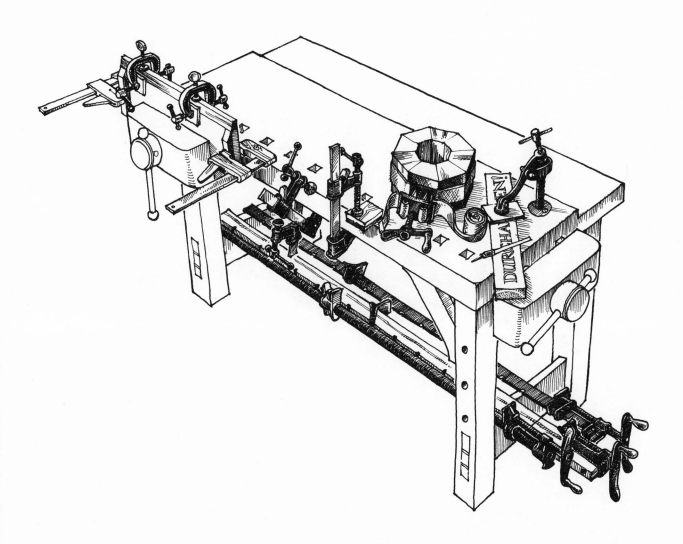

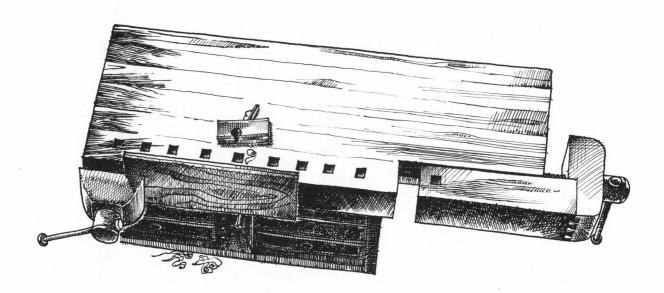

(*above*) A "German pattern" bench of the early 19th century
(*below*) One of many Shaker patterns for woodworking benches

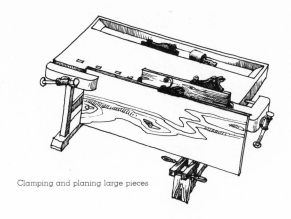

Clamping and planing large pieces

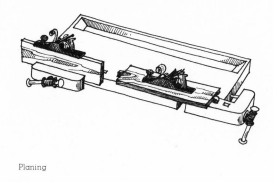

Planing

DISTRACTION, FATIGUE, AND SAFETY

There is one talent, one only: attention span. The ability to focus one's attention for a productive time makes possible all the feats of memory, music, art, science, mathematics, and craftsmanship. Everything else is mere preference. That a mind can reduce the constant intrusion of noise, random memories, tactile sensations, fatigue, and stomach rumblings to a single thread of directed concentration is an amazing ability, and to spin out that thread over a useful time is a gift.

Making something useful of wood is not easy. Making it useful and beautiful is doubly difficult. Anything you can do to reduce extraneous clutter and capture elusive concentration will help you.

There are two kinds of fatigue. One is the familiar drained-out feeling of bone-weariness. The other is subtler and often mistaken for a part of general exhaustion: it is the knotted muscle, the strained eye, the tense jaw that results from minor discomfort or an awkward position. This tension fatigue is a pilot for exhaustion and hastens overall fatigue by a factor.

In woodworking this secondary fatigue is a problem and can even be a danger. Accidents with power tools are largely caused by loose concentration, a loss of focus on the job and on the hands doing it. That concentration so important to safety is eroded by strained efforts at positioning your body or your hands in relation to the work, by noise, flying chips and dust, and by bad lighting.

You can improve your work and your chances of staying with a ten-digit system by seeing to your own comfort. It's not an indulgence; it's good sense. Your work area must be well lit, shadowless, and free from glare, especially around power tools. A face shield will ease the tension and distraction of squinting against flying debris, and a pair of ear stopples or ear protectors will lessen the annoyance of loud saws and whining routers. A respirator that fits well enough to wear comfortably can protect your fingers and your work against a sneeze at the wrong time, and it can also protect your lungs. Some wood dusts — from coca bola, pao ferro, Jamaican dogwood, and many others — are demonstrably toxic, but it's plain that no dust is a health treat for your lungs. Hang your face shield, ear protectors, respirator, and leather gloves (for handling rough lumber) next to your shop apron and *use* them. They can improve your work more than any gadget.

a Pulmosan respirator

a face-shield

In the first moment you see a Hammerli free match pistol you know it has affected the way you will see. It is an elegant solution to a design problem, the joining of the warm hand on one end to the cold precision on the other. From swelling curves of cherry that surround the shooter's hand the pistol changes to

Tools and the New Craftsman

by Ernie Conover

One of the most frequent questions asked by visitors to our shop is how we got started in business. This is not such an easy question to answer and takes some thinking. When I was discharged from the army in 1971, I went to work for a large automotive agency. I had been a sports car buff since I was old enough to drive, but dealing with the everyday problems of cars soon proved more than I could stand and my lifelong love of automobiles turned to hatred. The automobile is the primary reason for today's consumer rebellion. It is sold at an inflated price, and the average customer can't really trust his dealer. The ramifications of the energy crunch and, I suppose, changing attitudes on my own part, led me to the belief that a car was no more than a necessary evil.

I also felt stagnant at my automotive job as there was very little chance to be creative. I was approaching thirty and had a tremendous feeling that I had to get on with whatever I was going to do in life. After much contemplation and a good amount of prayer, I decided I could not devote my life to a dinosaur like the automobile.

I am fortunate to come from a creative fam-

ily. Both my mother and father are fine craftsmen in their own right. My father is an accomplished machinist as well as being an engineer. one of my happiest memories is from my twelfth year when my father bought me an English model maker's metal-working lathe. I still have it today: it is a Zyto made by E. Tyzack & Son and sports a 7-inch swing with a 14-inch gap bed. In addition there is a nice little vertical slide that allows it to be used as a milling machine. I keep it in a corner of the shop and use it now and again when there is a small piece of work and I am feeling nostalgic.

In 1975 the woodworking renaissance was in full swing and I was puttering in the basement like everyone else. At that time my father and I would get together about once a week for an evening in his shop. Often these sessions entailed the construction of tools that we needed for our woodworking. Among our projects was a set of threadboxes, which are taps and dyes for wood. Other things we made were knives, planes, and chisels. Our reasons for making our own tools varied. Sometimes it was because a tool was no longer available as a manufactured item. Other times, as in the case of some chisels, it was quicker and easier to

perfectly machined flat surfaces leading to the aggressive muzzle and sight, from the human to the inanimate. The same progression is repeated in a fine shotgun, from the figure of the Circassian walnut stock, checkered along the grip to converse with the hand and let in for the mechanism with patterned wood-to-steel

Conover thread box and tap set

joints that exclude a knife edge, to the abrupt brutality of the gaping bores and the terse brass bead sight. Even if you have never shot, the smoothness of the gunstocks speaks of their function.

make them ourselves than to go out and buy them. Finally, and most important, we often felt we could build a tool of much higher quality ourselves than could be obtained commercially. Added to this was the tremedous satisfaction of having done the job ourselves.

All this led me to the belief that there might be a place in the world for a small company devoted to manufacturing quality woodworking tools, tools sought by amateurs but not readily available anymore. With encouragement from my wife, father, and friends I set about building a series of prototype samples. I quit my job and went out on the road and tried to sell our ideas to various people. I say "our," because my father collaborated heavily with me during this period. I had a constant feeling of déjà vu; it was reminiscent of my high school years when I was constantly building science projects in the basement. In those times, too, my father was a constant source of encouragement, advice, and help. As in my high school years, my father seldom, if ever, put his hands to the work but rather brought the best out of me through a sort of Socratic method of asking very carefully worded questions. It's a technique that I suppose he honed to perfection in the running of an engineering department for many years.

When I went on the road trying to sell our ideas I was fortunate enough to go to the Brookstone Company. They were keenly interested in our prototype samples and asked us to quote on manufacturing some of them. Much of our success has to be credited to my wife's easygoing nature, because our dining-room table was turned into a drafting desk and the next month was spent in drawing out a series of threadboxes. This included some detailed work of calculating nonstandard thread geometry, and setting all the manufacturing tolerances. We built three samples from the prints and went back to Brookstone.

My trip to Brookstone was delayed three days by what now seems laughable. I had spent no small amount of time in building the three prototype samples. I had turned the bodies and machined and filed out all the metal parts. Finished, I set them next to my drafting board on the dining-room table for the family to admire. My wife suggested that we celebrate by going out to dinner. We left the newest addition to our family, a golden retriever puppy, at home. Apparently he took some exception to this, because by the time we returned home he had chewed one of the prototype threadboxes to smithereens. It was all my wife could do to restrain me from committing mayhem. Luckily, the metal parts were no match for his teeth and they were salvageable. I turned the mutilated wood parts for a second time and three days later I was off for Brookstone.

Brookstone issued us a substantial purchase order for some of our tools. Our first delivery was a mere six months away. We incorporated Conover Woodcraft Specialties under the State of Ohio and set about buying production machinery. I had an unused barn on my property and — with wiring and insulation — it became the headquarters for our company. It has served us well during the last three and a half years, and still has room for expansion. My father had just retired and was enthusiastic about the manufacture of woodworking tools. It didn't take much prodding to lure him out of retirement. We were in business and running. It is a credit to the patience of the Brookstone Company, because we missed our first delivery date by about six weeks.

When we founded Conover Woodcraft my father and I had some ideas that seemed rather radical at the time, and we weren't sure how they would work out in the long run. Like many consumers, we were often fed up with today's manufactured items, and had often thought, "Wouldn't this item be far better if they'd just spend a few extra cents here, or another minute or two there to make the thing really right?" We were just sick and tired of the cheap plastic part that failed or the permanently lubricated bearing that seized. (Many of *my* feelings on this matter came from having dealt with automobiles on a daily basis.) We decided to dedicate ourselves, as far as was humanly possible, to making a quality product, and, to some extent, letting the price fall where it may. That is, we try to bring a product to the marketplace at the best possible price, but if it takes an extra ten cents to buy a really high-quality component over a mediocre one, we tend toward the better end of the spectrum. In these years of high and rapidly rising prices, we had some doubts that we could live with this idea in the long run.

As it turns out, there are a large number of craftspeople who feel the same way we do and *are* willing to pay a bit extra to get something that is really well made. One of our best compliments came early in the game when we had an open house for the finishing of our barn. A man attending the open house walked up to me and said, "You know, all my life I've always thought that I would be happy to pay a bit extra if somebody would just build something of really high quality. I guess it's now time for me to dig deep down in my pocket." Needless to say we were floating on air for the rest of the evening.

In retrospect I am very thankful that the Lord led me in the direction we have chosen. At times it seems a bit futile and one can't help wondering if all the money wouldn't be better invested in U.S. Savings Bonds. On the other hand, there is no replacement for the personal satisfaction I get from the creativity involved in our business.

One aspect of work which I consider very important is personal growth. I think it is necessary, every so often, to review one's job and ask whether anything has been learned in the last day, week, or year. I am fortunate because my work allows me constant growth. There is no lack of engineering projects, as we are always trying to expand our product line. In addition there is a tremendous variety of work. A work day can be spent fixing a broken-down piece of machinery, working at the drawing board, or writing a letter to a customer.

Probably the greatest area of personal growth comes, however, from our customers. Each day brings us in contact with new and interesting people who are using our tools in creative ways. Customers will often write, tel-

The vise-grip is an indispensable tool. There are times when nothing else will do. It should be sealed in tombs and time capsules and carved into church decoration as an icon of our age. It has a sinister side, though, and as a matter of pure tool use the vise-grip is, like patriotism, the last refuge of a scoundrel. The vise-grip mentality has displaced a hundred tools that accomplished one task extremely well, and of whole families of hand-tools all we have are antiques in cases. What of the tools that fit one hand, the personally made, personally used tool? Gone. Yet there seems to be a recent reuse of many old hand tools, and more interest in the clever technology of the simple shape.

Country Craft Tools, a British book, can be more than a discourse on antiques. There are useful, beautiful tools in this plain book. The section on "Holding and Handling" has some ingenious old ideas for cramps (British clamps) and vises for any workshop. There is delight here, too. For me, in the appendix of craft names: reading through "fellmongers," "knappers," "lipworkers," "saw doctors," and "whitesmith" is like reading through the roster of a lost and honored regiment.

ephone, or stop in for advice on one of our tools. I find it interesting that during the conversation we frequently learn more from the customer than we impart to him. Woodworkers tend to be tremendously talented, interesting people.

This contact with the customer has led me to the firm conviction that craftsmanship is being kept alive by the amateur, though there are many professional woodworkers who are very talented and they certainly contribute to the art. It is a hopeful sign that people are increasingly willing to pay for custom work at a rate proportional to its worth. In these days of mass production I think it is still difficult for craftsmen to get a proper wage. The amateur has no such restraint hanging over his head. Because he does not have to sell his work he can concentrate on craftsmanship. There is no doubt that woodworking was in danger of becoming

a lost art form and, in this respect, there has been a true renaissance created by amateurs in the home workshops of America. We take great pride and derive satisfaction from supplying these people with tools.

It is the dedicated woodworker who takes pride in craftsmanship whom we consider when we design a tool. We try to take into account his needs and design a tool that will become an extension of his craft. Our tools are not for everyone: they are not intended for the person who just works to get a job done, but rather for the craftsman who wants a fine tool built with thought for form as well as for function.

All in all I feel very fortunate in my choice of vocation. Constant contact with diverse and stimulating people, varied and interesting work, and the chance to be creative. Few jobs can offer all of this.

Country Craft Tools could be a companion to three books by one surviving member of that lost regiment, Alexander Weygers. He was trained (in Holland) as a ship's engineer, a profession that demanded intimate acquaintance with every branch of toolmaking, tool use, and repair, due to the lack of hardware stores in midocean. He is a sculptor, a teacher, and an author, whose books *The Making of Tools*, *The Modern Blacksmith*, and *The Recycling, Use & Repairs of Tools* are well written, beautifully illustrated, and buoyantly positive. They all have a high "I could do *that*" factor, and are good reference for any shop.

Percy W. Blandford　　　　　$13.00
Country Craft Tools　　　　　240 pp.
Newton Abbot, England: David and　　Illustrated
Charles, Ltd., 1974

Alexander Weygers　　　$5.95 (paperback)
The Making of Tools　　　96 pp.
New York: Van Nostrand Reinhold,　　Illustrated
1973

Alexander Weygers　　　$8.95
The Modern Blacksmith　　96 pp.
New York: Van Nostrand Reinhold, 1974　　Illustrated

Ernie Conover, editor　　　$14.95
The Tool Catalogue　　　288 pp.
New York: Harper and Row, 1978　　Illustrated

If you have all the tools you will ever need and do not expect them to wear or break, you do not need this book. If your situation is somewhat different, I believe you need it. Under the impressive authority of *Consumer's Guide* every common and many uncommon tools are rated by manufacturer and model. The information includes performance ratings and prices, and in some instances can mean a better tool at a savings. It's a basic reference that will surely pay for itself in quality if not plain value.

119

Conover palm plane and scraping plane

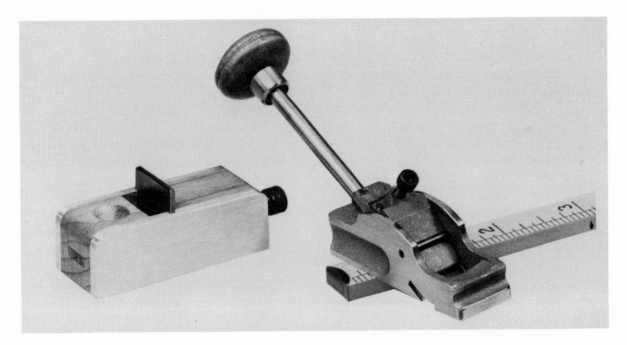

Ernie Conover likes his work and is proud of it. He has a boyish, infectious glee at doing what he wants to do and making it pay. He makes his tools as well as they can be made, and buyers find him. It's all very simple and very pretty and Ernie is a happy man.

A Whitehall pulling boat, gleaming in paint and varnish, is a pinnacle of craft in boatbuilding. Eight or ten species of wood assemble in a shape so whole that it begs to cut the water and glide over it. The Whitehall, as a piece of sculpture, is extraordinary, lyric, varied, fascinating, but without the hope of water, a dead thing.

Finishes

by John Ingalls

A number of years ago I had a job remodeling an old-time saloon in southeastern Alaska. Most of my ideas about how it should be done came from old movies. The trim was fir. I had gone to great trouble to buy pieces with the most beautiful grain and to make my joints immaculate. Naturally, I wanted this effort to show. My boss had different ideas. She was intent on staining the whole thing dark mahogany. At the beginning of the job I had given them the option of using mahogany and they said no. In protest I walked off the job.

They finally backed down and said that the color was still under negotiation. I went back to work but was still scared. For the next six weeks every evening after work I was at home trying to come up with a stain that would be acceptable to all of us. I tried everything from Clairol to mixing aniline dye with mashed banana. I finally satisfied myself by mixing aniline dye with shellac and applying it as a french polish. A similar situation had caused a friend of mine to give up woodworking entirely. In my case this experience was the inspiration for a lot of research, which has served as the backbone of this article.

Preparing the Surface

Sanding. Every craft has its own particular form of drudgery. Sanding is probably the least rewarding and most boring aspect of woodworking. To open an article with a section about sanding is equivalent to committing journalistic suicide.

As I perfected my skill in woodworking I gave up the disc sander in favor of the belt sander. Now I am finding that there is usually no substitute for hand sanding and scraping. What you don't know about sanding can be learned from a hardware dealer. The best I can do is give you some interesting things to think about while you are sanding.

Sanding grits are sized by agitating them in a tub of water. Sixty grit takes sixty seconds to settle; three hundred grit takes three hundred seconds to fall. One of the greatest improvements in sandpaper has been an electrostatic process, which causes a sliver of abrasive material to stand on end rather than fall on its side. The electrostatic sandpaper has many more sharp edges exposed. What more can be said except that you should sand with the grain

An ash-handled axe, an olive wood spoon, a big joint in a timber-framed barn, a hickory pack basket: their beauty grows out of their purpose and takes a life of its own.

Working on this book I have been trying to discover the wellsprings of craft. I have been hunting for the convergence of streams to follow them back to beginnings. I have seen

Sandpaper

conventional

electrostatically charged

and that sanding is an occupation ideally suited to faraway lovers on distant shores.

Scraping. Most of us have used hook scrapers to remove paint. For removing varnish from fine furniture a piece of broken window glass is the ideal tool. I have convinced a number of people of my insanity by searching the ground around building foundations to find a piece of glass the right shape. If you don't want to use naturally broken glass, a wheel glass cutter works perfectly. Remember that the wheel from the cutter ruins one side of each piece. It offers better control but wastes half of the glass it cuts.

Precision scraping. For the cabinetmaker broken glass does not offer a flat enough surface for his job. Rectangular steel blanks having the thickness of a saw blade are sold for this purpose. Sharpening this tool involves two steps. First, the edge is squared by holding it in a vise between two wooden blocks. The blocks act as a guide for an oilstone, which is rubbed over the surface. The blade is then drawn ½ inch above the blocks and a burnishing tool is drawn across the corner with great authority and pressure. This process exudes a sharp hardened burr the entire length of the

Forming the edge of a cabinet scraper

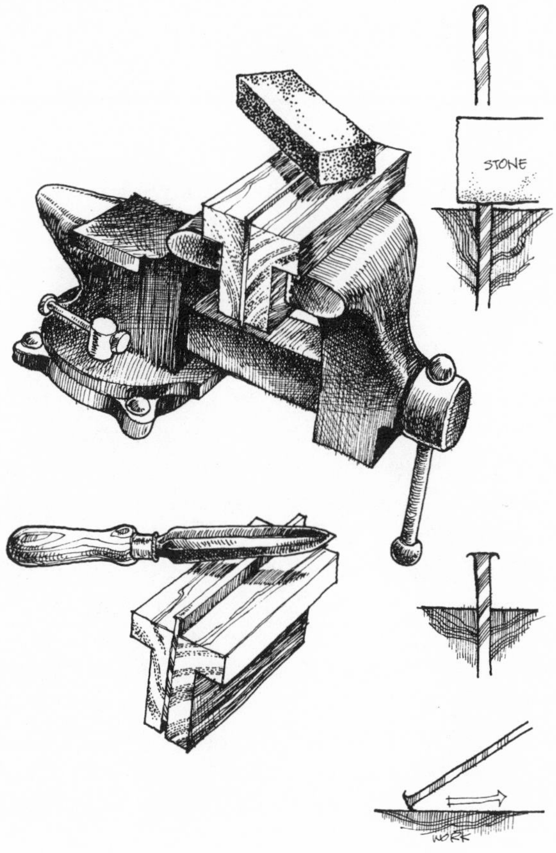

scraper. This burr is very effective for leveling a wood surface.

Stains

Our modern age seems to have bred three types of people: the double-knit disco group who like anything plastic; people who are happy to have anything so long as it is made

barns and instrument cases and marine blocks and sculpture and books and tools. Why is it, I have asked craftspeople and artists, that working with your hands is so important? What happens between you and your medium that satisfies you? Why is it important now?

of wood; and the bearded hipoisie who like everything to be au naturel. The combination of types has made it difficult to buy any quality stains.

I would like to make the distinction between dye and pigment. Dye is coloring material dissolved in solvent. The better the dye, the clearer the solution. Pigment does not dissolve but remains suspended in the vehicle.

Oil stains. There is a large variety of oil stains. The vehicle can be linseed oil, turpentine, mineral spirits, or creosote for outdoor work. Oil stains, at least those which usually come from the hardware store, can best be described as a turbid muddy paste in which the coloring is not thoroughly dissolved. These pigmented stains are made by grinding powdered pigment into oil. This is an easy process and most of the stains on the market are produced in this way. The colors are quite permanent and they are easy to use. I recommend such stains if you want to cover up the natural grain of the wood with a muddy film. If the can says "stir before using" you know you are buying pigmented oil stain.

The production of a dye oil stain requires a good deal of research. It is difficult to find dyes that will actually dissolve in oil. The coloring materials are derived from anilines and other coal tar derivatives. The colors thus produced are more fugitive than water stains but they are easier to use. Dye stains do not obscure the natural grain of the wood, making the results quite tolerable to the eye. In a city of 100,000 you would be lucky to find one paint store carrying this type of stain. To my knowledge the only firm that makes dye stain for wood is McClosky in Philadelphia. Unless you are a chemical engineer, I expect that you will be using a commercial product and there is no need for me to give directions. Let me only say that pleasing effects can be gotten by starting with light-colored dyes and finishing off with darker ones. In ash, for example, it gives the wood a rich hue and fills the pores with dark contrast.

Water stains. For the expert woodworker, water stains are the most satisfactory. These stains come in the form of powdered aniline dye to be dissolved in hot water. All the colors can be derived from red, yellow, and blue. If you are ordering material, I suggest buying basic brown, adding red, yellow, and blue to give the exact color you want. Anilines are difficult to buy and I have often resorted to buying small packages of clothing dye like Rit.

Water stains have a tendency to raise the grain of the wood. This problem can be avoided by first sanding the piece and then mopping the entire surface with a wet sponge to raise the wood grain. Resanding will again flatten the grain and it will not be lifted by the stain.

Modern anilines are highly stable and being insoluble in oil will not bleed into the finish.

Alcohol stains. Aniline dyes are also soluble in alcohol. If the dye was intended for cloth, it contains salt normally used as a mordant to fix the dye. These salt crystals will not dissolve in

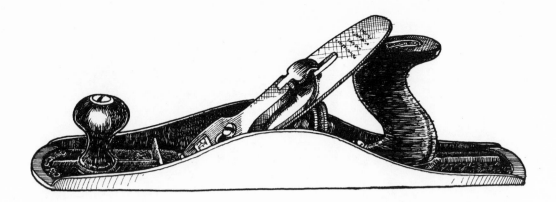

alcohol and must be strained out of the solution.

Alcohol stains have all the advantages of wood stain except that the solvent is expensive and dries so quickly that the results tend to be splotchy. Alcohol does not raise the grain as does water.

Colored shellac and varnish. Certain woods, most notably fir, spruce, and, to a lesser extent, pine, are made up of alternate bands of pith and hardwood. The pith is white and the hard grain is light brown. Penetrating stain applied to these woods is repelled by the harder bundles and absorbed by the pith. The natural grain of the wood is reversed and the result might be compared to the face of a woman who wears mascara on her lips and lipstick on her eyelashes. The woodworker who chooses to ignore this fact is little more than a hack and will be destined to a life of customizing vans and building surfboards.

Stradivari, Guarneri, Amati, and other seventeenth-century fiddle makers solved this problem by first sealing the wood with clear varnish then adding numerous coats of dyed varnish. Pigmented varnishes are easily gotten commercially or by adding paint pigments to varnish. The results look like glazed mud. If you enjoy working evenings in your basement with such ingredients as dragon's blood, gum arabic, and Venice turpentine, perhaps you can rediscover the secrets of the old masters and make a varnish-soluble dye.

After numerous failures I discovered a practical and faithful method of staining these rib-bon-grained woods. The surface is sealed with a fine coating of shellac. Dissolve aniline dye of the desired color in shellac. Strain to remove the salt crystals. Apply the colored shellac by brush, spray, or french polishing. The same effect can be obtained by applying a sealer coat of lacquer and following it up with numerous layers of dyed lacquer. This technique is followed by modern violin makers. Sunbursts can be created by starting with yellow then covering it with successive darker layers. The areas that have been rubbed down with pumice then take on a lighter appearance.

Chemical stains. Certain woods, notably oak and chestnut, contain large amounts of tannic acid. In the presence of ammonia fumes the tannic acid gives the wood a beautiful brown hue resembling tanned leather. The wood is placed overnight in a sealed box or tent containing several pans of household ammonia. After the wood has reached the desired shade, a coating of boiled linseed oil will preserve the tan.

Grain Filler

Grain filler is more of a convenience than a necessity to the woodworker. The purpose is to fill the pores of open-grained wood. The same effect can be obtained by sanding and varnishing the surface two or three times. Various materials including cornstarch flour, clay, and plaster have been used with limited success to provide a body for the boiled linseed oil base.

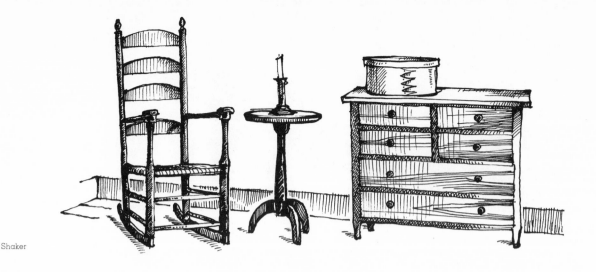

Shaker

124

The most satisfactory material is ground flint, called silex. The very sharp corners of this material bind it together and also lock it into the wood pores.

The basic technique is to drive the paste deeply into the pores, brushing with the grain. Brushing across the grain will build up the surface. Allow the paste to harden for several hours and wipe with a rag, working across the grain.

Oil Finishes

Drying oils have room in their chemical structure to absorb oxygen. In doing so the film polymerizes, forms a hard surface, and a piece of wood so coated gains weight as it dries. If you open a can of paint and feel a vacuum as you lift the top it contains drying oil which has absorbed the oxygen out of the can. This drying process is similar to combustion. The same property that makes these oils polymerize also makes them spontaneous-combustion hazards.

Dryers. Oils are composed of large organic molecules. The rate at which oxygen can penetrate from the surface to the inner molecules is normally very slow. Tiny metallic salt molecules like P10 and MnO_2 have the ability to absorb oxygen at the surface and carry it back to the inner molecules. Perhaps the situation is analogous to a bicycle messenger carrying a dispatch through heavy traffic.

Linseed oil. The old standby of the paint and varnish industry is linseed oil. Its ability to absorb oxygen out of the air is greatly increased by boiling or the addition of dryers. The surface thus produced is fairly hard and moderately impervious to water. Linseed oil is not highly recommended for light-colored wood because it tends to yellow and the surface never becomes hard enough to repel dirt. On boats the surface coated with linseed oil acts as a host for mildew.

Tung oil. Tung oil has proven to be the Saran Wrap of the oil industry. It has stronger drying properties, is harder and more water repellant, and forms a clearer shield than any of the other drying oils. Marco Polo described the trees with their heart-shaped leaves, from which Tung oil is derived, on his travels in China. For centuries the Chinese have gathered these nuts, removed the husks, and pressed out this marvelous oil. They used it to treat the stones of the China Wall and this treatment has accounted for its preservation. Production of Tung oil in the United States started in Gainesville, Florida, in 1932.

Danish or Watco oil. This oil is actually not an oil at all. These products contain phenolic resins dissolved in a slow-evaporating mineral solvent. Phenolic might be compared to violin rosin but it is much more durable. These penetrating finishes sink into the wood and the solvents flush off, leaving the pores full of a hardened resin. Many of these products contain a small amount of oil to temper the resins and to facilitate the rubbing.

Lemon oil. Lemon oil is not a true finish in that it doesn't harden like other oils. Applied on bare wood it has little value except to prevent the absorption of moisture. The acid from the lemon acts as a cleaning agent and the oil tends to hide scratches in a lacquered or otherwise finished surface.

Application of oil finish. According to old wives' tales the proper way to oil a surface is once an hour for a day, once a day for a week, once a week for a year, and once a year for a lifetime. While this treatment would yield a good-looking piece of furniture, it is not really necessary and goes against good reasoning.

The goal is to fill the wood pores with solid material. The first step is to get the oil to penetrate as deeply as possible. This is done by heating the oil, thinning, and keeping the surface drenched as long as possible. After four hours or longer every trace of oil is removed so that the oil on the surface does not prevent the inner oil from hardening. The process is repeated two or three times, with thicker and thicker oil as we work from the inside out. The final layer of oil is not wiped away for perhaps eighteen hours, until it has begun to polymerize. Certain cabinetmakers actually allow a scum to form on the surface and then rub it down with turpentine and steel wool.

Wax. The use of wax for polishing furniture dates back to the 1920s. It was during the boom of the movie industry that a Washington decorator published an article showing pictures of highly varnished furniture having mirrors glued to the surface. Americans were becoming affluent enough to express their bad taste and the fad caught on. Now everyone could live like a movie star. Wax was a cheap, easy way to get the desired shine. The problem with wax is that it remains soft and acts as a trap for airborne dust. Years of rewaxing produce a black scum on the surface that is sticky to the touch and can be scraped off with the fingernail. Wax offers little resistance to moisture and the wet glass put down on a waxed surface usually leaves a ring. If you insist on using wax, I suggest removal between successive coats.

Varnish

In the ancient Greek city of Cyrene lived a beautiful woman named Berenice, who had long amber-blond hair. She was the faithful wife of Ptolemy Euergetes, king of Egypt, who was away in battle. Soon after their marriage Berenice vowed to sacrifice her golden hair on the altar of Aphrodite should her husband come back safely from war in Asia. Her shining hair covered with jewels was wafted from the altar to heaven to form a constellation, *Coma Berenices*. From this story of the beautiful hair upon the altar comes the word varnish.

Varnish includes any of a number of clear transparent finishes divided into two cate-

gories: spirit varnish and oil varnish. Varnishes have the obvious value of protecting and beautifying wood. They also defend furniture against the forces caused by poor design. Varnish seals the wood and in doing so it stops the hygroscopic expansion and contraction of opposing grain.

Spirit varnish. Spirit varnish is produced by dissolving a gum or a resin in a solvent such as turpentine, alcohol, or acetone. During the drying process after application this solvent evaporates, leaving a hard protective film.

Dammar varnish. Dammar varnish is produced by dissolving dammar resin in turpentine. Dammar resin is slightly harder than rosin and crystal clear. The fact that it contains no drying oil also prevents it from turning yellow. Dammar varnish is used for coating paintings and other objects where whiteness is a goal. It also forms a body for certain white enamel paints.

Shellac. Of all the finishing materials shellac has the most interesting history. The raw material is produced by the lac bug whose name is derived from the ancient Sanskrit word *laksha*, meaning 10,000. During swarming time thousands of these tiny insects attach themselves to the branches of various types of fig trees. The lac bug sucks the sap of the tree and encases itself in a reddish-colored resin. With a multitude of insects this resin develops to a thickness of up to half an inch and encrusts the small branches and twigs.

The villagers go out, collect these twigs, called stick lac, and scrape off the resin. The powdered resin is bright red and until the development of anilines was a source of dye. This color is removed by washing with hot water. The remaining powder is stuffed into long cotton bags the diameter of a baseball bat. Heating and twisting causes the resin to ooze out of the bag. The addition of .05 percent arsenic sulfide and 5 percent rosin facilitates this process. The molten lac is placed on the side of a large earthenware jar and finally stretched. The stretched sheet is shattered to form flake shellac.

> To make varnish you must have Spirit of Wine, which must be strong, or it will spoyl the varnish, and not dissolve your Gums, and consequently hinder your design; for the stronger your Spirits the better will the varnish be. Therefore the best way to prove your spirits, is to take some in a spoon, and put a little gunpowder in it, and set the spirit on fire . . . and if it burn so long till it fire the gunpowder before it go out, it is fit for use, and will dissolve your gums. *

Commercial shellac is produced by adding three to four pounds of flake shellac to one gallon of denatured alcohol. The proper mixture for finishing work is usually two pounds per gallon. Alcohol evaporates so quickly that applying shellac to a surface is a difficult problem. It can be brushed very quickly or sprayed but the most artful solution to the problem is french polishing.

* *A Treatise of Japanning, Varnishing, and Gilding* (1688).

French polishing. The housewife in olden days usually devoted one day each week to making sure that all the furniture was properly oiled and polished. At the international exhibition of 1851 a new method of finishing was on display. The method eliminated the need for this weekly drudgery and came to be known as french polishing.

French polish produces the most exquisite of all finishes. It has depth and luster without having the vulgar appearance of high-gloss oil varnish or the contrived look of matte varnish. French polish stands up well against time but like the face of a handsome woman is quickly aged in a tavern or a drinking establishment. The two biggest enemies of french polish are the thoughtless smoker and the otherwise harmless drunk who happens to put down his glass on the finished surface.

Shellac solvents dry so quickly that the brush is not anxious to slide across the surface. French polishing is an ingenious solution to this problem. The brush is replaced by a linen pad, as shown in the pictures. To keep the pad from sticking, a small amount of linseed oil is used as a lubricant. This process is similar to spit-shining shoes.

I have found lots of variation in the advice given by practitioners of french polishing. I will say categorically that you should avoid any premixed preparation. Quite often the piece is rubbed with boiled linseed oil and allowed to dry for several days. This will darken the final product.

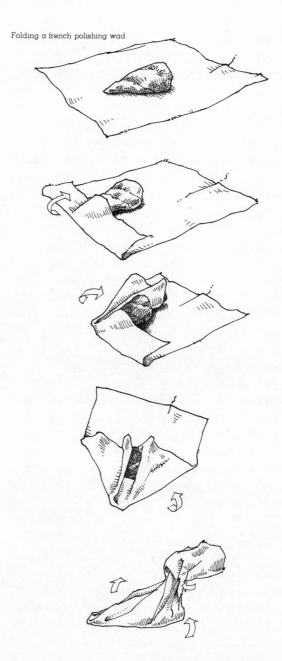

Folding a french polishing wad

Sizing. Sizing is accomplished by rubbing one to two pounds of shellac on the entire surface. The rubber should contain no oil. Any oil will be selectively absorbed by the surface and produce a splotchy appearance. If you like, you can repeat this process several times with light sanding in between. I do not suggest sanding after the final sizing as it may cause the open grain to absorb the oil out of the rubber. Sizing is best accomplished with long determined strokes with the grain.

Bodying in. Bodying in refers to the building up of a lot of layers of shellac. This is accomplished by wetting the rubber to the extent that drops form when it is squeezed. Several drops of linseed oil are placed on the bottom of the piece and circular strokes are used, as shown in the diagram. If darker tones are desired colored shellac or aniline can be used.

Spiriting off. Spiriting off removes the excess oil and burnishes the final surface to make it smooth. Add a few drops of alcohol to a new rubber. Squeeze it and touch it to your face. It should feel cold but without the hint of any liquid. Spiriting off is best accomplished using large figure-eight strokes.

Oil Varnish

Going into a hardware store and asking for a can of oil varnish is like going into a grocery store and asking for a dozen hen eggs. The salesperson looks perplexed and goes to the backroom to speak to the owner. Finally he returns with a can and a guilty look on his face, never quite sure that he is selling you what you want.

It can be argued that varnish making reached its prime in about 1700. This art was developed by the Italian fiddle makers.

> How fascinating is the appearance of the varnish as now seen in fine examples of Stradivari instruments! Lightness of texture and transparency combined with brilliant yet subdued colouring, and above all, the broken-up surface more especially of the back, form a whole which is picturesque and attractive in the highest degree.*

By the year 1784 all the violin makers started using spirit varnish in order to save time and money. During the intervening years until the present, the formulas for making oil varnish have disappeared and it has become a lost art. There is little doubt that this varnish contributed greatly to the excellent sound of these instruments.

According to recent accounts, Stradivari, before he died, copied the formula along with application instructions in the family Bible. This book was destroyed, however. Signor Giacomo Stradivari apparently made a faithful copy of the recipe which he jealously guarded. He has been asked on numerous occasions to produce the formula and this is his answer:

* Alfred Hill, *Antonio Stradivari* (New York: Dover, 1963).

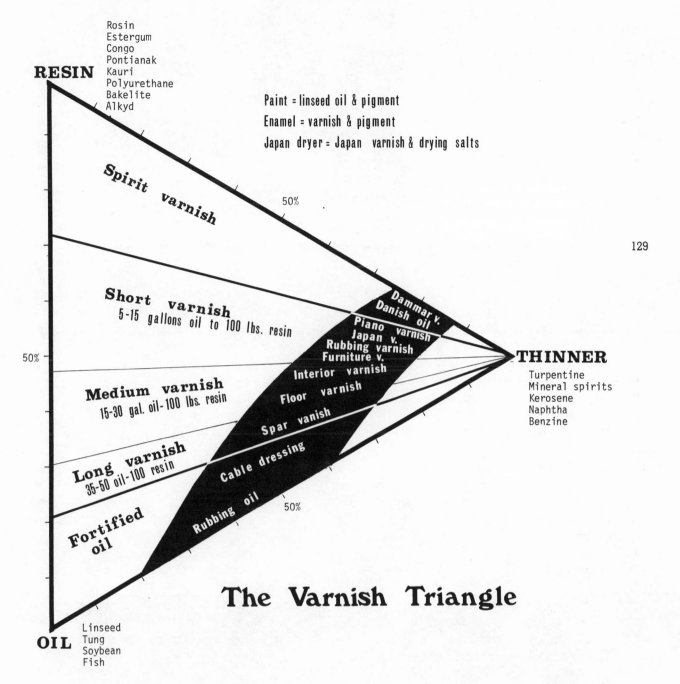

RESIN

Rosin
Estergum
Congo
Pontianak
Kauri
Polyurethane
Bakelite
Alkyd

Paint = linseed oil & pigment
Enamel = varnish & pigment
Japan dryer = Japan varnish & drying salts

Spirit varnish

50%

Short varnish
5-15 gallons oil to 100 lbs. resin

Dammar v.
Danish oil
Piano varnish
Japan v.
Rubbing varnish
Furniture v.
Interior varnish
Floor varnish
Spar vanish
Cable dressing
Rubbing oil

50%

Medium varnish
15-30 gal. oil-100 lbs. resin

Long varnish
35-50 oil-100 resin

Fortified oil

50%

THINNER

Turpentine
Mineral spirits
Kerosene
Naphtha
Benzine

OIL

Linseed
Tung
Soybean
Fish

The Varnish Triangle

129

"You make an impossible request, one which I cannot grant you as I have never confided the secret of the varnish even to my wife or my daughters. You may consider it an eccentricity on my part but nevertheless until I arrive at a different opinion, I wish to be consistent with and remain faithful to the resolution of my youth never to reveal to anybody the contents of this precious recipe. If by chance other Stradivaris — my sons, nephews, grandsons or grand-nephews — should turn their attention to mechanics, more especially the craft of their celebrated ancestor, they should at least have the advantage of possessing the recipe of his varnish."

Varnish is composed of three main ingredients: resin, oil, and thinner. By altering the concentration of these three ingredients the properties of the varnish can be changed. Varnish short on oil tends to be hard, brittle, and very clear. It can be rubbed with fine abrasives such as pumice and rottenstone and brought to a high polish. As oil is added to the mixture the varnish becomes softer, more durable, and more weather resistant.

The resin is first heated in large kettles. At a certain critical time and temperature the drying oil is added and cooked for about

Chippendale

130

twenty-four hours. Drying salts are added and the mixture is thinned.

The composition of varnish is best illustrated by a triangular diagram. As we can see the percentage of oil varies from 30 to 85 percent. We also find that varnishes higher in resin require much more thinner.

Japan dryer. Japan dryer is actually a varnish having a high resin content. To this have been added large quantities of drying salts — MnO_2, PbO, Pb_3O_4. For an explanation of how these drying salts work, read the dryer section under oil finishes.

Resins

Rosin is used extensively in cheaper varnishes. It produces a clear initial finish but quickly breaks down under the force of time. The primary source of rosin in the United States is the southern longleaf pine.

Ester gum is produced by cooking rosin and glycerine. The result is a resin much more durable and usable outdoors.

Pontianak is a hard resin highly suited to spar varnish.

Congo is a hard resin ideal for furniture and other applications where rubbing is involved.

Polyurethane stands up well to abrasive wear. For this reason it is excellent for floors. Its powers of adhesion are not quite as good as other resins.

Bakelite, being highly chemical resistant, is used for making spar varnish. It has a ten-

dency to yellow and is used in conjunction with a material that acts as an ultraviolet shield.

Alkyd is an inexpensive resin having many good qualities.

Drying Oils

Tung oil produces the best varnish. It is clear, rot resistant, and hard. Tung oil is not very elastic and is often mixed with linseed oil for outdoor use.

Linseed oil is less expensive and less desirable than tung oil. It is more elastic but tends to yellow.

Soybean oil is a weak drying oil and does not perform well without the use of chemical dryers.

Fish oil does not weather exceptionally well. It has a great ability to penetrate into rust and for this reason is used as a base for rustproofing primers.

Thinners. Some of the thinners used in varnish include turpentine, mineral spirits, naphtha benzine and kerosene. Although seldom used in industry, turpentine is certainly the best. The fussy painter always uses turpentine. It has wonderful solvency and tends to speed the oxygen molecules into the varnish and aids in drying. Turpentine is too expensive to be used for cleaning brushes.

Turpentine is two kinds. Gum turpentine is produced in the United States by tapping the longleaf pine. The sap is distilled to make a whole range of products including turpentine.

Distilled spirits are produced by heating stumps and other resinous parts of the pine tree. The vapors are again distilled and turpentine is one of the products. Generally gum turpentine smells better and is of better quality.

Enamel. Enamel is a type of varnish to which pigment has been added. For this reason enamel is very glossy and durable. Paint in the true sense merely consists of linseed oil and pigment.

On varnishing. In the United States the leading practitioners of varnishing are yachtsmen. This fussy and overzealous bunch of people on the East Coast have turned varnishing into a sport and on the West Coast into a religion. It is here that the author of this article is in grave danger of punishment for unsportsmanlike behavior — nay, heresy.

The chemicals in wood have a tendency to inhibit varnish from drying. The application of a shellac primer makes the varnishing process go more quickly. This procedure is not recommended where the piece is exposed to water. In this case the first coat of varnish should be thinned with one-quarter turpentine. (Thinning is best done twenty-four hours before varnishing.) The first layer of varnish should be very thin and I myself apply it with a rag.

I usually apply varnish with the grain, then brush across the grain and then with the grain again. On vertical surfaces I always end with vertical strokes. This keeps the varnish from running.

The hardening of varnish involves two steps: first, the solvents evaporate; then the drying oils absorb oxygen and polymerize. This exchange with the air is greatly slowed down during damp weather.

Many of the yachtsmen use sandpaper that is much finer than necessary for sanding between coats. I have found that anything from 150 to 200 works fine. If you desire a rubbed finish, dip a rubbing pad in linseed, paraffin oil, or water. Sprinkle pumice on the surface and rub with the grain. For highly polished surfaces this is followed by rottenstone.

Varnish removal. All varnish must someday be removed. The ideal varnish is one that is durable but that can someday be removed with a minimum of effort. Many a piece of furniture has been ruined by encasing it in a miracle plastic that will never come off.

Varnish remover contains a number of volatile solvents. The most effective removers contain a wax that floats to the surface and keeps these solvents from evaporating. Every trace of this wax must be removed with turpentine before revarnishing. Waxless varnish removers do not require this cleaning but are far less effective.

Lacquer

The use of lacquer dates back to 500 B.C. in China and Japan where they slashed the bark and collected the sap of the lac tree. This resin was dissolved in camphor, and various oils (tung oil, etc.) were added to temper the resulting product. The lacquer thus produced was certainly better than anything else of its day and possibly better than our modern varnishes. These Oriental lacquers dried by oxidation and moisture hastened this process.

The lacquers of today are best thought of as fortified spirit varnish. They contain the following ingredients:

1. *Nitrocellulose* or nitrated cotton is produced by dissolving cotton in a combination of nitric and sulfuric acid. In its chemistry it is similar to the highly explosive gun cotton. After being dried nitrocellulose is soluble in ethyl alcohol, anhydrous ethyl acetate, and other well-known solvents.

2. *Resins* are added to lacquer to lower the cost, soften the film, and increase adhesion. The resins commonly used in the past have been shellac, kauri, and dammar. Today's lacquer usually contains alkyd resin.

3. *Solvents.* Nitrocellulose and resin are not soluble in one solvent. Therefore lacquers contain several compatible solvents. Common examples are ethyl ether and ethyl alcohol, ethyl alcohol and ethyl acetate.

4. *Thinners.* The solvents used in making lacquer are expensive. For this reason the solution is thinned with a different material such as ethyl alcohol or naphtha or benzene. Most lacquer will only accept a certain amount of thinner without having the nitrocellulose drop out of solution.

5. *Plasticizers and softeners* are added to lacquer to soften the film, increase adhesion, and make brushing possible. Castor oil, cam-phor, and tung oil are most commonly added to lacquer.

Lacquers have become very popular for finishing musical instruments and furniture. They are easily applied with a spray rig, have a wide variety of dyed colors, and can be easily rubbed.

Paint

Had I no love for mankind I would put down my pen. My social conscience, however, demands that I say a few words about paint. The Leonardos of today have created a number of miracles of dubious social value. The most obvious of these is nuclear energy. Also falling into this category is latex paint.

In the days when painters came to work wearing ties, the goal was to spread the thinnest, most uniform coating possible on the wall. This called for thorough preparation, washing and making the wall a uniform color before painting. Where a wet brush or roller overlaps dry paint, the double coverage makes that spot stand out. To avoid this problem painters used slow-drying paint and as many as six painters might work down a wall at one time.

The paint of that time was fugitive and, having the consistency of honey, tended to creep down the wall. It is no wonder, especially considering the number of people working in close proximity, that painters have gained a reputation for being alcoholics. Modern technology

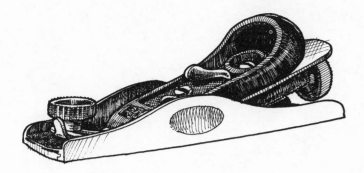

has created latex paint that is thixotropic. Instead of being sticky like old paint, it is highly docile to the brush. Like mayonnaise, the paints have the ability to spread and stay put and the modern painter is no longer competing with gravity. These mayonnaise paints cover up a multitude of sins and now we are living in a world of self-proclaimed experts on painting.

I would like to take this opportunity to point out some sobering facts on the subject. Each time we paint, the room becomes smaller. The rate of buildup with latex paint is about one-fifth of an inch per century. At that rate by the year 2500 our doors will be three and one half inches instead of one and a half inches thick. By the year 10,000, walls will have one and a half feet of paint on them and we will have to breed a whole new generation of people short enough to keep from hitting the ceiling.

The moral is, before you get out a brush, think. Does the wall really need paint, or can it be washed? Many buildings in New York City are being ruined by improper painting. I have even seen paint stalactites hanging from pipes. Before you start to paint get a little sad and remember that someday it will all have to come off.

References

Painting and Decorating Encyclopedia (South Holland, Ill.: Goodheart Wilcox Co.).
  A great book for the painter.

Newell and Holtrop, *Coloring, Finishing and Painting Wood* (Peoria, Ill.: Chas. A. Bennett Co., Inc. 1961).
  Well written and informative.

A Treatise of Japanning, Varnishing and Gilding (1688) (Chicago: Quadrangle Books, 1960 reprint).

Hill, Alfred. *Antonio Stradivari His Life and Work, 1644–1737* (New York: Dover, 1963).

John Ingalls is an extraordinarily curious man. He has a quizzical mind, a restless nature, and a background in geology and physics that have made him more attuned to the warp and woof of nature than anyone else I know. He is a journeyman in many professions, from tree-trimming to machinist and marine architecture. He can frame, roof, wire, dry wall, glaze, and make exquisite cabinetry, when the need arises. He has great, stained hands that do not suggest grace and yet the films he has made for young people are almost lyric and his renaissance and baroque flutes grace the softer hands of flautists here and in Europe and enrich the collection of (among others) the Smithsonian Institution. I do not know what he will do next, but I haven't known for fifteen years. Presently he is part owner of the Crystal Saloon & Ballroom in Juneau, Alaska.

The Cabinetmaker and Commissions

by Dean Torges

It's a challenge earning a living as a cabinetmaker in the Midwest. Most people think of a cabinetmaker as a craftsman who makes kitchen cabinets or restores antiques. I've done both, and a good share of plywood built-ins, lodge furniture, display cases, speaker cabinets, and more, but this work requires only competent skills and is undertaken to keep creditors' campfires from scorching our porch. Commissions for original furniture are the reason I'm a cabinetmaker.

They come by word of mouth and begin as open business propositions. But a special kind of mutual respect develops between the patron and the cabinetmaker, a respect based on the creation from wood of an object that both helped birth and for which both share affection. My patrons visit while the work is in progress and become knowledgeable in its particulars. We stimulate each other, and the excitement continues after the piece is finished. I still write to and attend an occasional baseball game with Paul and Corita Kollman, who commissioned my first dining room, even though we live over a hundred miles apart.

There must be much trust for a commission to work fruitfully because we aren't choosing from catalogue pictures and booking shipping dates. Usually the piece we decide to do changes in the process of construction. Ideas come at odd moments in odd ways and one must be loose and secure enough to follow up on them. They hide in the margins of experience and get discovered accidentally. Sometimes I have a very exact idea of what needs to be done in a particular area, and may even be working from my own scale drawing, but perhaps a lathe tool slips, or a critical cut is wrong, or maybe it just looks like hell in the wood and flesh. I used to stomp around like Rumpelstiltskin and throw it across the room, but I've learned to sit back for a moment and think about it because there's often an alternative to retracing footsteps. It seems that some dark part of us sometimes knows about a better path than our consciousness, and if we're respectful we can bring it to light and come up with something significantly better than we'd planned. This has happened to me often enough that I am almost reverent about the whole process and have developed a new kind of patience with myself. It's a humbling admission to say that mistakes and idiocies have proved a most fertile source of personal inspi-

136

ration, yet they are, after all, just the flip side of the coin. You grow from them if you look around, rather than curse your clumsiness or stupidity.

A more orthodox source of ideas is the work of other cabinetmakers. At first, I felt some shame relying on the creative accomplishments of others, but I see now that such theft comprises the greater part of creativity and that its euphemisms are "influence" and even "inspiration," words that tend to distance us from the act of creativity by suggesting mysterious or divine sources. But creativity is not a force bounding about, fueled by mystical rushes of innovation; it inches forward and involves stealing a little from this guy and a little from that guy and adding one's own angles, thereby "inspiring" the cabinetmaker down the road to larceny if everything blends successfully. That's the way it goes, in furniture, in literature, sculpture, painting, whatever. In the process, the artist takes enough to be identified with the community of man, and adds enough to make himself a proud lifetime member.

My biggest problem early on concerned pricing my furniture. In this respect I must have been like any other craftsman or artist who cared to such an extent about the quality of his work that he invested his identity in it. The risk was not the loss of a sale that frightened me,

but the vulnerability to a sort of personal rejection; and the tendency, therefore, was to sell cheap. Second, time sheets are okay as long as one is turning out picture frames or kitchen cabinets; prices thereby are arrived at fairly and certainly. But one cannot be concerned with time and simultaneously undertake the creation of a dining-room table and chairs that represent the very best one can do in design and execution and be able to say at its completion, "That's my best shot. I wouldn't change anything." Time, in such circumstances, is an enervating spectator whispering persuasive arguments for compromise. And if one plugs one's ears to such arguments, the tendency again is to sell cheap because there are few people willing to fund what is ultimately a very private pursuit, a quest for excellence. Yet that's exactly what patrons do. Clients, on the other hand, are satisfied with an adequate shot in the direction of excellence. One must either educate clients into patrons or, failing that, absorb the difference.

Yet a fair price must be arrived at, not in the name of economic justice, but because any craftsman or artist exploring through his work can only do so fearlessly when he finds the bravery to say, "Here I am in this work, and if you see a sympathetic reflection, it costs *X*. If you don't, it's only because of the diverse roads

each of us travels." Bargains, in such a circumstance, are like compromises; they may make good economic or political sense, but they retard the spirit.

Production furniture has reached its nadir, both aesthetically and intellectually. Its trend now is to see if it can crowd a four-by-four member into the place of a one-by-one, affecting the appearance of strength and permanence. This is a practiced deception that seeks to satisfy both the brags of consumerism and the needs of productivity. Its premise works from the outside in, presuming that external appearances alone matter, so it asks, "What's the most economical way of holding it together on the inside without allowing this cheapness to bleed through and affect the appearance of an outside uncompromisingly crafted from timbers?" The result of this premise is that words like "cosmetic," "facade," and "veneer" have earned distasteful implications in the language, that they never before deserved.

This is the chute we've slid on since the Industrial Revolution, when the designer, the cabinetmaker, and the business director emerged as three separate people. The arrangement is necessarily antagonistic and can only produce consistently deteriorating products, for each office is powerless to control the product absolutely, and therefore no one is willing to invest himself in his work because of the probability of someone else smudging his fingerprints. Without this personal check and as an act of private preservation, each man withdraws into the safety of anonymity and they close ranks in a headlong rush to-

 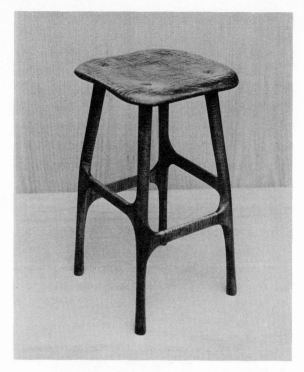

138

ward slovenliness. To paraphrase an American critic who witnessed this process a century ago, a corporation of conscientious men can exist, but not a corporation with a conscience.

However, in a grass-roots movement across the nation the three identities are being refused as one person, as they were in the Golden Age of American furniture making, and in this sense we are experiencing another revolution, a renaissance. Exploration and reevaluation characterize it rather than a recognizable furniture style. We see everything from massive stacked sculpted pieces to wispy flowing laminated ones, reproductions built with period tools to gnarled busy masses that are perhaps faithful reflections of the chaos this part of the century lives with. Within this turmoil, directions for the twenty-first century are brooding. If we share anything in common, it's that most of us lack business sense. That's okay for now. Michelangelo may have

struck a bad business deal, but he mixed his own paints.

If I had to characterize my own furniture, I'd do it with the word "simplicity." In defiance of consumerism, I want its brag to be "Look how little I need." Indeed, in my mind that must be the brag of the future lest we eat ourselves off this leaf of earth. What's the least needed to build the greatest strength? To me, the two ideas of strength and simplicity are complementary, not contradictory, and they don't entail austerity. I want furniture to be intelligible, for the outside and the inside to be fused so that the reality is the appearance; for its integrity, in the simple mathematical sense of integer: indivisibility, to convert an observer of it into a participant in it. I don't want someone to look at it and marvel at what I did, but instead to marvel at what can be done, what we can all do, and not just with wood.

Cincinnatus was ploughing his field outside Rome when the city fathers found him and begged him to leave his oxen. They wanted him to save Rome. My fantasy is that Washington will helicopter into Ostrander, Ohio, and ask Dean Torges to leave his shop and four daughters and garden and his lovely, gentle wife, Mary. Goodness knows we need saving, and Dean — who is a philosopher as well as a writer and craftsman — might well be the man to do it, but I think he will stay in Ostrander. He is too happy where he is and his batting average is improving.

This is what I think happened. I think that we became more specialized as the world around us grew more sophisticated. It became impossible for each person to make his own goods. We moved away from our imperatives — food, shelter, a chair, a spoon — and bent our worry and sweat to other needs. Still, craft was part of everyone's life: if you did not make it yourself you asked an artisan to make it,

Some Thoughts on Cabinetmaking and Hand Methods

by Tom Clark

Furniture makers who want to be anything more than producers of potentially useful objects ought, I think, to dismiss the notion that the end result justifies the means in their work.

There are plenty of furniture makers who are content to build utilitarian wooden objects — tables, chairs, chests, and the like. They are not worried about anything other than making furniture that functions as it is supposed to and that looks like they want it to. These makers do not philosophically care very much how they get their piece together because their mental energies are solely directed toward the finished product. They want to use it, or see it used and admired. They are fixated on the results of their work.

Most of the tastemakers and critics of non-industrial furniture making are themselves furniture designers or builders or both. They want to base their judgments on a furniture piece unencumbered by ticklish distinctions about how the final result came to exist. Rather than having to qualify their opinions with an analysis of demonstrated personal skills in a work, they find that it is easier to disregard this aspect altogether. It is an unpopular business to offend peers and contemporaries.

The resulting overemphasis on result has generated the current prevailing critical philosophy that, prima facie, equates simplicity with effectiveness, complexity with pretentiousness, and that champions ingeniously contrived mechanical process over technically difficult human process. The elimination of personal skill as an obvious factor in evaluative critique has promoted simplistic designs and illogical constructions, often tortuously mechanical in method.

I submit that it is absolutely necessary to take into account methods of manufacture to assess validly the quality and importance of furniture. The questions of what tools and what skills have been applied must be answered. The furniture maker, in setting his own goals, must understand that there are qualitative differences between classes of work, in difficulties of construction, and in sophistication of execution. To believe that consideration of process is unimportant in critical opinion is to misunderstand the differences in human endeavor. If a desk, after all, is just a desk, then any suitable platform is the *bureau du roi* in artistic terms, and I do not believe it. To negate the importance of the means to an end is, for me,

just so, to fit your house or your horse or your foot. In some times and places survival needs were so abundantly fulfilled that objects became images, they assumed a ceremonial importance quite apart from their original function: a war club lost its lethal function and became a jeweled scepter, and a suit of court armor was admired for its art more than its protection against broadswords and arrows.

to miss the central motivation in unessential making of items in the first place.

As a cabinetmaker, I am a practitioner of an obsolete trade, one long ago superseded by corporate enterprise and mass production. Cabinetmakers are simply not needed to make furniture as furniture. If all makers of individually designed and built furniture disappeared tomorrow, they would not be missed, nor would their absence be telegraphed by a lack of household furniture.

Therefore, I consider myself an artist, because that is the only justification I have for doing what I do. It is easier to rationalize obsolescence, I find, by knowing that painters still paint, even though there are other ways to make pictures more technically proficient and prolific.

I approach the making of furniture using the traditional and customary practices developed throughout the eighteenth and nineteenth centuries in true hand workshops. I start my furniture from rough-sawn wood and carry through the whole piece in the trade workman's way with hand tools, most of them antique.

I do not use power tools to make furniture because I consider them inappropriate for building individual pieces of fine furniture. The standard range of machinery used today is part of mass-production technology. If I use it in the making of my furniture, what I wind up building, in effect, is mass-produced furniture one piece at a time. What little time can be saved with machines is insignificant considering the thousands of factory pieces spit out

in the same time it takes to build an item. In any event, the machine is seldom used in modern workshops to save labor, but is used, in the final analysis, as a replacement for skill. The furniture maker winds up being an industrial designer who thinks he is a craftsman because he is his own machine operator. His furniture is designed around, and built with, tools he can use without needing a body of refined manual skill.

In the twentieth century there is not much artistic or practical point in worshiping furniture that can be, perhaps ought to be, supplied by the mechanized industry. Most of what is built by individual furniture makers today falls into that category. Each man or woman has a limited working lifetime, and since time is a necessity in producing any piece of furniture, taken as a body of work, there is not a whole lot one person, or group of people, can produce from the starting shot to the finish line. To make items that fail to stretch the builder's talent; to make items that do not develop refined personal skill; to make items that do not extend the artist to the outer limits of his, or her, abilities is a sad misuse of time.

I believe it is wrong to assert that the new man-made materials, machines, and fabrication methods are in and of themselves reasons to abandon the artistry of true hand work. I can sympathize with the heartfelt urge creative people have to be new and different. I can understand wanting to raise a new icon to design and construction. But in the face of the almost unattainable level of skill shown in the past, craftsmen cry "it's been done" to avoid

As the need to struggle for life became more distant, the roots of function became more obscure and design changed. A sideboard or highboy was once used to keep plates, napkins, and dinnerware clean in a house heated by open fires, floored with rushes or chips, inhabited by working dogs, farmers, and the occasional bird. Later, these functional necessities became decorative pieces, serving more as

being measured against that venerable yardstick. It is safer to be judged by a new stick struck with one's own increments.

I wonder how many people do not carry out certain processes saying that they are not interested, or that the techniques are foolish and laborious, when, in fact, these builders lack the skills to undertake the work. Without the traditional manual skills the designer and mechanic are lock-stepped into a progression of industrial choices, which for practical purposes eliminates the artistry I believe only true handwork can demonstrate.

Knowing what I know about the history of furniture and the long traditions of the trade, I am concerned about cabinetmaking as a fine art. I do not like the term "designer-craftsman" because I feel it is an apologetic term. It is a puffed-up label used by furniture makers, and others, who are insecure about calling themselves artists. Many rightfully so. I believe there is no valid reason other than artistry to be a professional furniture maker in the late twentieth century.

Tom Clark is the Master Cabinetmaker and Conservator of Furniture at the Ohio State Historical Center in Columbus, Ohio. His training in stage and theater design must have included table saws and electric drills but now he is seldom asked to endorse power tools.

A British chair bodger

"WOODWORKING 101"

One man's palm plane is another's power plane and it is difficult to recommend one reference on carpentry and woodworking over another: some have minor faults; some fall short here but excel there. The illusions of democracy and literary truth make for a lot of word-squirming, yet there are a few cautions and suggestions we can make without too much waffling.

Decide what you want: carpentry, cabinetry, or a specialized side pursuit like turning or finishing. Do you want a shop reference for specific data or general information or a guide to learning or plans for projects? Is your level of skill past primers, past fundamentals, past general knowledge? What do you want to do: repairs around your house or apartment, home improvement, simple carpentry, barns and outbuildings, home building, simple cabinetry, or fine joinery? Are you going to work with hand tools, hand power tools, shop tools? Do you have a good supply of Band-Aids?

Jan Adkins $4.95
Toolchest   48 pp.
New York: Walker & Co., 1973   Illustrated

This is a primer, dealing with basics. It's an introduction to woodwork for young people or for beginners of any age. The step-by-step illustrations demonstrate the first skills of measuring, marking, cutting, chiseling, planing, and gluing. I'm partial to the book because I wrote and illustrated it, and I wonder how Krenov ever became so good without it.

Leo P. McDonnell $8.95
The Use of Hand Woodworking Tools   273 pp.
New York: Van Nostrand Reinhold, 1978   Illustrated

This seems to be an updated edition of an old shop-class classic, and though the tests at the ends of the chapters give me the jitters ("Didja pass, Mo?") it's a good basic work with clear photos, simple illustrations, and readable text. That's pretty good.

vehicles for craft than as furniture. Probably no more skillful cabinetry was ever produced in the Western world, but it was little more than ostentation. Skill outstripped function in the fantastic inlaid patterns ("seaweed," Chinese, landscape, and flower motifs), the *bombé* shapes and rippling fronts, the excesses of turnery and hardware, the lengthy conspiracies to hide every structural joint. The cabi-

Charles H. Hayward
Practical Woodwork
Buchanan, N.Y.: Emerson Books, 1967

$5.95
192 pp.
Illustrated

Carpentry for Beginners
Buchanan, N.Y.: Emerson Books, 1975

$6.95
200 pp.
Illustrated

Together, these small British books present a quantity of tidbits on working wood by hand, some of them not found in more recent manuals. There is a genuine flavor of the workshop here, a feeling of well-worn tools in the slightly dusty explanations and the charming prewar photos. The woodworker shown has massive hands with serving-spoon thumbs and the effect is comforting, encouraging. I like these books.

Robert Campbell and N. H. Mager
How to Work with Tools and Wood
New York: Stanley Tools Edition, 1952;
Pocket Books, 1975

$1.95
488 pp.
Illustrated

There is a lot of information in this book for its size. Its illustrations are small (it is a standard-size paperback) and poorly printed on paper of minimum quality, but generally clear with an accompanying text that should not confuse anyone beyond his powers to return. The "shop projects" at the end of the book are fairly mindless, but the whole presentation is fair. If lifeboats had workshops, this small book would be a good addition to the toolkit.

Fred W. Zimmerman
Exploring Woodworking — Basic Fundamentals
South Holland, Ill.: Goodheart-Willcox, 1976

$6.65
208 pp.
Illustrated

This is a textbook, complete with quizzes at ends of the chapters and projects suitable for the folks. Nevertheless, it has clear illustrations and good photos and an exhaustively clear text at something below a sixth-grade level. The designer and typesetter should be chased around the publishing house with a stick.

Mike Bubel and Nancy Bubel
Working Wood
Emmaus Pa.: Rodale Press, 1977

$3.95
192 pp.
Illustrated

The raison d'être for this bit of country folk is that all that old used wood lying about the spread is going to waste, so why not roll up your homespuns and hike up your calicoes and get to work respecting it? The photographs herein are fine, and if you favor the country-funk style you'll be partial to Ms. Liz Buell's illustrations, which are clear, if earnestly unlabored, but the instruction is slight and primary and, sometimes, goldanged frightening. The Bubels' log-round garage walls will eventually crush — at least — some domestic fowl . . . and probably the pickup, if not the driver. There seem to be some workable goat-shed devices here, if that's your pleasure, but little you couldn't figure out on your own.

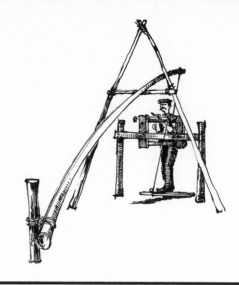

A bodger's pole lathe

144

Raphael Teller $7.95
Wood Work: A Basic Manual 142 pp.
Boston: Little, Brown, 1974 Illustrated

This is a strange piece of work of little use to anyone. It
has an expensive four-color jacket around a confused
presentation of stock photos, inept illustrations, and almost
no sense of continuity between the tools and the finished
products (which are, incidentally, simple and inoffensive).
The tools for this "basic manual" were chosen by a curious
process — beside a hammer and a nailset is a six-foot
table jointer; a hand drill appears beside a massive drill
press with a mortise attachment. Chairs, the badge of a
woodworker's best efforts, are given a scant page and a
half of text and several photos of finished chairs. A statement by Mr. Teller, not a book, this should have explained
his skill in another way.

Intermediate fare:

R. J. De Cristoforo $13.95
Complete Book of Power Tools 434 pp.
New York: Harper & Row, 1972 Illustrated

A fine book, thorough and explicit with excellent photographs and illustrations. Plain basics and fancy techniques. De Cristoforo is especially strong on the use of jigs
for safe and easy handling of special operations. If you
intend to use power tools this book should be part of your
shop.
Another De Cristoforo book, *Power Tool Woodworking
for Everyone,* is more a shop manual for the Shopsmith
machine, a power tool that can be set up to saw, drill,
sand, lathe, cut, chop, dice, and fricassee. It's obviously
a formidable piece of equipment but the resetting time
between operations seems inconvenient. In any case, Mr.
De Cristoforo's instructions are just as useful here.

John L. Feirer $27.50
Cabinetmaking and Millwork 992 pp.
Peoria, Ill.: Chas. A. Bennett Co., 1967, Illustrated
1970, 1977

A large, useful shop manual with a lot of information
about power shop tools and set up, this book is heavily
illustrated with good photos and line drawings. It's a good,
comprehensive shop reference at a hefty price.

Jeannette T. Adams $8.95
Arco's New Complete Woodworking 752 pp.
Handbook Illustrated
New York: Arco, 1975

Fairly comprehensive, this is a useful book but not the
only book you'd ever want on woodworking. As an author
I grit my teeth at any book that purports to be "complete."
Its examples and illustrations are understandable and its
coverage is good. Some of its copy is lifted directly from
other books, but if it's valid information don't be snobbish
about it. It features a strong first section on hand tools and
moves on from there.

netmaker's mandate was to counterfeit a pretty world held together by wishes, a world without savage needs. He built artifacts from a world of the mind and never revealed the structure below the facade. These pieces show us the consummate skill of the craftsman guided by the whim of fashion. Many of them have survived to attest to the structure beneath, but they have outlived the time of ser-

Carpentry is a deep subject to plumb with one book, but you might as well start somewhere:

Willis Wagner $12.80
Modern Carpentry   480 pp.
South Holland, Ill.: Goodheart-Willcox, Illustrated
1976

Straight, simple, clear, and broad, with good photos, good illustrations — this is a textbook, but you could learn a lot. A Munchkin figure yclept "Handy" turns up every few pages to recommend that you follow the manufacturer's instructions or to suggest that you not clean your teeth with a saber saw but with a little white paint (semigloss enamel); you can expunge "Handy" and the chapter tests and have yourself a good reference.

J. Douglas Wilson $2.95
Practical House Carpentry   424 pp.
New York: McGraw-Hill, 1973   Illustrated

While rather a basic presentation, this certainly contains a formidable amount of information.

Harry F. Ulrey $7.50
Carpentry and Building   440 pp.
Indianapolis: Theodore Audel and Co.,   Illustrated
1966, second edition 1978

The Audel Guides are usually good reference, but this particular volume suffers from a question-and-answer format, making it less valuable than other books reviewed here. I suspect that one or more of the four volumes in the Audel Carpenter's and Builder's Library would be of more use: Volume 1, *Tools* (steel square, saw-filing, joinery, cabinets); Volume 2, *Math* (plans, specs, estimates); Volume 3 (house and roof framing, laying out, foundations); Volume 4 (doors, windows, stairs, millwork, painting). The old Audel *Carpenters and Builders Guides* often turn up under a tool tray in your uncle's shop. Take them. They are full of information and fine engravings of old tools you've never seen, and they have sections on using wood planes and other reviving tools. Write Audel for a list of their other guides.

Jeffrey Ehrlich and Marc Mannheimer $7.95
The Carpenter's Manifesto   318 pp.
New York: Holt, Rinehart and Winston,   Illustrated
1977

Any book that claims to "take the mystery out of carpentry for everyone" is to be regarded with braced skepticism and baleful looks. The deepest mystery in carpentry resolves itself with a little head scratching and some work. The idea that carpentry is a can of corn once you cross through the mysterious veil demeans a very straightforward profession that happens to require a lot of learning, a lot of details, and a lot of technique. I do not believe you can learn from this funky, inept play for the browser's dough.

vants and daily care. They seem curious to us now, alien.

A body of work that seems closer to our world is the Shaker style of building and cabinetry. We can puzzle over their beliefs but it was that stern and pervasive philosophy, an overspreading idea of simplicity that applied equally to the soul and to the physical world, that freed them from ornate fashion. In our

Peter S. Stamberg
Instant Furniture — Low Cost, Easy-to-Assemble Tables, Chairs, Couches, Beds, Desks, and Storage Systems
New York: Van Nostrand Reinhold, 1976

$7.95
160 pp.
Illustrated

There is a manic appeal in this book and in its designs. The pieces photographed and detailed (very nicely) here look like third-year design-school projects, like the furnishings of a contemporary Robinson Crusoe, or like the utilitarian comforts of a bridge-construction shed. The standard construction lumber flaunts its knots and its twelve-point crosscut ends, but do not reject this book out of hand: many of its designs are genuinely innovative and, unlike most furniture construction, take structural advantage of triangulation. For a beginning woodworker or an open-minded fine joiner there is something here to learn.

The Editors of *Organic Gardening & Farming*
Build It Better Yourself
Emmaus, Pa.: Rodale Press, 1977

$16.95
942 pp.
Illustrated

This is the best and the widest in scale of the "project" books we've seen. As such it should be a stimulant to idle tools. The several hundred house and farm and garden and barn and workshop projects shown here are explained so well, illustrated in such a lucid, clear way, and designed with such simple strength that no beginning wood butcher, no experienced but reluctant carpenter need be intimidated back into the living room. Plans for wheelbarrows, composters, all-purpose carts, arbors, cold frames, gates, small sheds and barns, salvaged windows, greenhouses, sawhorses . . . This is not cabinetry nor even joinery, but it proves the beauty of simplest woodwork and tinkering around the place. It's got good sections on repairing old outbuildings, on building stone walls, and on frame construction. It's a book I'm glad I've got.

time a generally high standard of living makes a house with servants rare; a family usually sees to its own cooking and housework with the aid of electric servants and a simpler style of furnishings. Our tastes have grown nearer to the Shaker styles and we can admire the purity in their work that is reasonable and lasting.

Ken Braren and Roger Griffith
Home Made
Charlotte, Vt.: Garden Way Publishing, 1977

$6.95
176 pp.
Illustrated

This is not a good book of projects if you consider it an adult book, but it would make a good book for a youngster beginning to accumulate bruises and Band-Aids from learning woodwork. These useful projects are simple enough to encourage a beginner, with a high success rate at useful items.

Stu Campbell
Build Your Own Solar Water Heater
Charlotte, Vt.: Garden Way Publishing, 1978

$7.95
105 pp.
Illustrated

According to the Office of Science and Technology, about 15 percent of residential energy use goes to heating water, a substantial and costly bloc of our energy bill. This is an excellent guide to using the sun's power for most of that 15 percent instead of the power company's. There are workable water heaters here that you can build in an afternoon with a few dollars' worth of pipe and fittings. There are also high-tech active solar panels and the numbers for them. A good overview, this book gives enough detail to complete the job.

Some cabinetmakers today want to show their prowess with complex styles and claim that Shaker furniture is popular because it is simple and cheap to make. No compliment could have pleased a Shaker craftsperson more; it would have been a proof of Shaker belief. Their workshops were models of efficiency. They sold Shaker furniture and furnishings to the world outside their belief. They

148

A Baldwin Tub, and very inviting it is

devised labor-saving machines to keep costs and labor low, including the circular saw (invented by Sister Tabitha Babbitt in 1810). The Shaker goals were subtler than ostentation. Proportion, simplicity, and strength are more exacting yardsticks than complication and flamboyance, and more difficult to use. Filigree and ornate molding can be used to cover poor joints, cracks, and checks, and veneer cov-

John G. Shea $16.95
The American Shakers and Their Furniture   203 pp.
New York: Van Nostrand Reinhold, 1971   Illustrated

Well written, well designed, and illustrated with excellent photos of Shaker furniture and their surroundings, this book is especially successful in its evocation of the spare life of the Shaker community; this is due in part to Shea's skill, but just as much to the artifacts of Shaker life shown, speaking with their clean, quiet logic. One hundred and four pages of museum-measured drawings of Shaker "smallcraft," "utility designs," and "furniture classics," with construction details and photographs of the pieces, comprise a pool of inspiration for wood butcher or fine joiner. A highly recommended book.

Joseph Daniele $14.95
Building Early American Furniture   256 pp.
Harrisburg, Pa.: Stackpole Books, 1974,   Illustrated
second edition 1978

This is a book of simple early design approachable by most craftsmen. The explicit drawings are encouraging and uncluttered, and most designs are accompanied by photographs of the original pieces. An appealing book, it could keep a workshop gainfully busy for a long time.

Nancy A. Smith $5.95 (paperback)
Old Furniture: Understanding the   192 pp.
Craftsman's Art   Illustrated
Boston: Little, Brown, 1975

An impressive amount of information is available in this detailed appraisal of antique furniture. Through photographs and illustrations (the latter could have been better) the cabinetmaker can learn what damage time and use can give his work, the owner of antiques can learn to repair his pieces conscientiously, and the buyer can learn to recognize fakes. The book is well designed, well executed, and well written.

Kevin Callahan $5.95
Early American Furniture   158 pp.
New York: Drake, 1975   Illustrated

A book of agreeable drawings with hand-lettered text, brown ink on cream paper in an attractive format, this is aimed at antique collectors, with a section on refinishing. It's pleasant, but light. The basic style of the drawings and the tone of the text persuade me that this might be a good book for young readers interested in refinishing a piece of furniture or working with you in the shop.

Thomas Moser $6.95
How to Build Shaker Furniture   210 pp.
New York: Drake, 1977   Illustrated

If the photos in this book had been printed well, and if the illustrations had been drawn by a finer hand, this book would have been a handsome piece of work. Even now, the publisher's mistakes notwithstanding, it is a valuable book for its comfortable text, its comments on construction details, and its measured drawings of the photographed pieces.

Making books is very hard, a profession full of pitfalls and disappointment. I may take up piracy as more satisfying and healthier in the fresh-air line.

ers a multitude of sins. The plainer the piece, the more perfect the joints and the wood must be.

Though our tastes have become simpler, our lives have continued to diverge from the base line of living. We are so specialized that few of us can produce any of our own goods. The things we have are bought from an enormous stock of items designed for everyone

WILLI SUNDQUIST
by Lance Lee

Willi Sundquist. Woodcarver. Out of Sweden and out of a tradition profound in its directness and simplicity comes an inspiration in wood. To study under Willi for a week is to encounter an old and wonderful transition, a brief, enigmatic description of enlightenment: at first a mountain is a mountain and a river a river; after much study and meditation these become symbolic, shrouded, mystified; and in the end they become again, after enlightenment, a mountain and a river. But with what a difference!

To carve a spoon appears, at first, easy enough. First, one grapples with the subtleties of design, of wood selection, with the feebleness of the eye in attaining a pure oval, with the difficulty of transcending the ordinary and the humdrum in the design of the common spoon. Then one becomes aware of the symbolic potential, the enticements, the levels of technique unthought of, the leverage of thumbs as fulcrums, the optimism of the handle . . .

The spoon becomes imbued with complex, delightful significances, enshrouded with a faintly intimidating prospect. Beyond this mystic level of possibilities Willi himself, a carver of spoons, seems to have come through a dark and fascinating tunnel into a magical light in which the spoon is once more but a spoon, "but with what a difference."

Along the way much rubs off upon a student of Willi's. The awareness that all around lies fine wood suitable for bowls, toys, spoons. In a stack of firewood or a blown-down limb lies all that is needed. At first you believe that he carries all of the tools needful in his small haversack — a tiny curved adz, a hewing hatchet, a set of gouges, a pair of knives, an apron to ensure that his hostess will never despair of woodchips in the rug. Then one watches Willi run out a spoon blank on a bandsaw and finds something troubling, that he is after all *not* entirely independent of plug-in machinery. You mention it. Very shortly Willi has you hewing out a "blank" with a hatchet from a piece of birch in your own stove-woodpile and in a very few minutes Willi has liberated you. Not from thinking that *he* required power tools, but from believing that *you* did. Strange currents begin to move in you and you ask for a

little more, thank you. He shows you your spoon blank, very correct, the bowl shaped on the bottom, cut away, toed in at the end to save time later with the knife, the top happily level and the long, arching handle not at all unshapely, not at all bad. With the gentlest reproof he shows you your grayness, your mediocrity, how common a button-down mind you actually have. For you have been watching Willi's work, enjoying his design, the distinctive curves and hollows, the little protrusion or hook from which the later-to-be-hung-up spoon will show itself off in its unassuming fashion. But you entirely missed the point. Willi's handle turns *up*, a mischievous, grinning, positive handle which the thumb controls independent of the wrist and the grain has become exciting and stands out proud . . . a spoon of distinction and, yes, a spoon with a whole temperament and a philosophy. Yours looks like a fast-food franchise's sugar slopper. He takes a scrap of birch and illustrates the difference between a smiling, optimistic bowl or handle and the manufacture of negation which will endure within the material culture, as a result of *you*. How do you say thank you for such a gift?

His course in woodcarving is not truly a woodworking course. It is one in philosophy, masked by cherry and white pine, a piece of walnut which he found while walking along the street and favored bits from next February's woodpile. Wrong again. Willi offers something altogether closer to the human affair: a course without a name.

This man is quietly, steadily encouraging people to do things together. He is a teacher in the tradition of Leo, the servant-mentor in Hesse's great book, *Journey to the East.*

Of an evening people can sit and carve together, sharing their designs, the curves and the good grain as it emerges from the piece, brought up and released by tenderness, deliberate cuts by oil and tenderness. They may share a marriage of the practical and the beautiful as the hands fair down a section of the handle which requires little concentration natural in their task, automatic, and the conversation drifts away from wood, spoons, form, function, and they find themselves operating on two planes simultaneously, enjoyable and effectively. You are

... not for any one person, specifically, nor for any specific home, but for a general loose fit to indistinct needs. We fit ourselves by the nearest classification: a 16/32 short, a 9½D shoe. Our homes are from stock plans designed on familiar models, oriented more to the street than to the sun, furnished with rugs and chairs from a catalogue. We drive in machines brilliantly engineered to decay at a controlled

sitting beside a master craftsman and master teacher and watching his hands as they emulate him, talking all the while of religion or Spanish icons or the artistry of Carl Larsson, learning a most magnificent way. Ideas seem to merge with the exercise, the world of the intellect and the senses, the hands, the material realm, all merge and things are as they ought to be. No, this is not a course in philosophy; this is a way of life, an answer to riddles: "How do I bring up my kids?" "How can I duck down beneath the shroud of the superficial and the mundane?" Really, Willi has merely turned out a little wooden horse — earnest, patient, prepared for the lifetime ownership of a little boy, and I've only chipped out an imperfect circle, the business end of a maple spoon. Have I imagined the rest?

As a little boy I was given a small animal carved, so I thought, by a Transylvanian woodcarver. The little fellow cast a spell over me which I've never lost. A small book came with it, about the life of the woodcarver — the shavings, the snow, the early light on the mountains, the incredible air . . . about the romantic simplicity and satisfaction in which he lived. I don't recall whether he had family, a mortgage, rheumatism, was badgered by tourists or had shrill relatives. I recall that there was magic in the wood and in the hands and that when I grew up — tinker, tailor, insurance salesman or Bay City crane operator — one of the options would be that of woodcarver. Willi Sundquist is a woodcarver. He is alive and well, rich in those things that have made him the inspiring tender, generous man that he is. And his pockets are not empty.

Lance Lee is a tight, sputtering fuse of enthusiasm whose work at the Apprenticeshop of Bath, Maine, is perpetuating skills and attitudes that were in danger of extinction. The passing on of real technique and the communication of the satisfactions in handwork and wood knowledge is enormously important and this bright-eyed, excitable man carries it on like a quest.

rate along straight highways that disregard terrain, to jobs that add one piece to a puzzle too large for viewing. We are strangers to the seasons, too far away from the grit and scrabble of living to know winter as anything but inconvenient cold, spring as more than mud, and the song of summer is the hum and drip of the air conditioner.

I think we have been carried safely away

The Whittling Disease

by Jake Page

It's a damned contagion.

Your brother-in-law in Baltimore takes up carving ducks out of basswood, gives you a hunk of the stuff, and you get a simple Stanley knife at the hardware store and whack away at some fool idea you have in your head. You succeed in turning the idea into three dimensions without cutting the skin off your knuckles and without completely violating the grain. Basswood, you discover, is nice stuff — fine grained, light, somehow made to be responsive to the sharp edge of a knife.

Just then a subversive teacher at your preteenage daughters' school interests them in chess. With the chintzy allowance you give them, they buy a cheap plastic set. Being competitive, you buy a book about the game and keep its wisdom close to your chest while whipping your daughters at their own game.

In an idle moment, you look dreamily at the plastic bishop with his slotted hat and it reminds you of a shark. Why not carve a chess set? With the bishops as sharks? With one side all predators, the other side their prey? A nice confluence of your kids' interest in chess and your own in natural history, the kind of thing that's good for the whole family.

The next Saturday there is a shark bishop.

The Saturday after that there is a coiled cobra — a knight. The rooks emerge as vultures. You oil these basswood stalwarts and wax them. The other side, the prey, shall be dark. You will pit ferret pawns versus rabbits.

Ultimately the prey-predator scheme breaks down. Shark bishops slant against dolphin bishops. Vulture rooks swoop down on penguin rooks. The swift queen, a cheetah, is poised against a dark kangaroo queen. The predator king is that ultimate ecological badass, a human, staring across the board at his counterpart in the primate world — a pensive gorilla. Your human king's base is a bit small and he tends to wobble and fall down if not handled smartly. Never mind. Press on . . . pawns. Carving eight rabbits and eight ferrets gets to be a bit of a bore.

Being regularly employed and having certain unavoidable household obligations, you take a year to complete the set. Having spent your spare moments engaged in the existential concentration of forcing your ideas on the compliant basswood while your daughters have spent theirs playing chess, you lose the first game played with this zoological freak. And the second and the third. You lose interest in chess. The fun, after all, was in the carving.

from the dangerous part of life by currents of civilization and mechanization and specialization. I think we are losing our bearings and there is an uncomfortable feeling of disorientation that becomes stronger as we move further away. More of us search out the ridges of life where wind has scoured away cover at the thinnest places and exposed the rock needs beneath. This may be why we sail and climb

and backpack, and this may be why we build things.

My grandfather's house and my god-father's house were full of tinkering: cupboards to hold one object, swiveling phone stands, a toggle and rope disappearing into the ceiling that opened a ventilator on the third floor, inventive ways to store pots lids, a tiny shelf that held a thermometer at just the right

A saddlemaker using a drawknife against a breasthook

Your wife likes turtles. You lay hands on a sweet dark piece of walnut. Hard, resistant wood, the sheer beauty of which makes the work worth it. Your wrists grow strong and you get cocky, winding up with a thumb laid open by your fine new bench knife sharpened on Arkansas stone. After a trip to the local emergency room and three weeks of thumb recuperation, you warily return to the walnut and, a month late, you give your wife her birthday present, a walnut turtle barrette — a shell under which the hair-holding pin slides, a turtle's head mounted on a flexible stick fashioned from a chopstick.

A year or so later, a rambunctious young dog chews the barrette to flinders. Oh well, the fun of it was in the carving.

Mercantile matters take you to British Columbia one November. There, to the west of the charming city of Victoria, is a body of water called Esquimalt Lagoon and you and your wife, recently become avid birdwatchers, see twenty-three species of ducks and grebes while standing in one place in the cold rain. A mind-riveting spectacle, and your knife hand begins to twitch.

Over the months, little basswood ducks begin to appear on the mantelpiece — about two inches long, painted with gouache: you will re-create the entire spectacle in miniature, a pair of each species you saw on Esquimalt Lagoon. You reach eighteen species by about the next Christmas. The spirit of the season overcomes your normal churlishness and a rather large part of the duck population winds up in the hands of relatives and friends who

(bless them) have admired your spectacle.

A remainder vanishes into the pocket of a Hopi Indian gentleman whose sons, he explains, are of certain clans associated with the remaining ducks. The mini-spectacle is gone. Ah well, the fun was in the carving. And in knowing that your modest creations, the quiet fiddling of hands, wood, and blades during idle moments, are being enjoyed by friends here and there across the country, friends you are not very assiduous about writing.

You get interested in egrets with their marvelously sinuous necks. Your wife takes your first egret (cherry wood) and puts it on the highest (dog-proof) shelf of the bookcase, having extracted a vow that you won't give it away to anyone ever for any reason.

You feel good.

You buy a set of chisels and a Marple mallet. You order one of those fancy German woodworker's benches.

It's a damned contagion.

James K. Page, Jr. is a tall, lanky fellow with quiet eyes and a sly grin. When he is eighty-five he will still be saddled with the adjective "boyish." His principal talents seem to be relaxing on low chairs in a draping, pythonish way and whistling Jack Teagarden jazz, but this is subterfuge; his real talent is supporting the illusion of calm over his own boiling imagination and a great, leaky house filled by his beautiful wife Suzanne (a noted photographer), six young women (noted daughters), and various hangers-on (noted eaters). He is a distinguished editor, most recently for Smithsonian Books, a passable writer, and the gentlest, most generous soul I know. He and Suzanne spend their time outside Washington, D.C., with the birds and with the Hopi, two groups who know friends when they see them.

height, a mirror fixed to reflect a view of the driveway to the occupant of a favorite chair. When my grandfather heard a car the satisfaction at seeing it without stirring his brittle legs belonged to him, he had invented that satisfaction. His twisted fingers had shaped some part of his own comfort. I remember them as old men, tinkering with gnarled, rebellious hands over small, clever things. They

One Highly Evolved Tool Box

by Jay Baldwin

Our portable shop has been evolving for about fifteen years now. There's nothing really very special about it except that a continuing process of removing obsolete or inadequate tools and replacing them with better ones has resulted in a collection that makes trying out ideas uncommonly easy. It's a generalist's shop; we're not equipped to produce fine cabinetry. We can work any common material. It is orderly enough to permit others to find things, but it stops short of constricting, anal neatness. It enables things to get done with less hassle, and so more gets done with less hassle. We use it as a three-dimensional sketchpad.

We mostly shop Silvo and U.S. General. Their prices nearly always beat local stores even when you add postage. Sometimes the savings are dramatic. You should compare these catalogues with each other and with Sears to get an idea of what you are saving. There's nothing more annoying than sending away for something that turns out to be cheaper a block from your home. The mail-order places have a larger selection than local outlets. More esoteric tools can be found in the always-tempting offerings of Brookstone, Gar-

rett-Wade, and other specialty houses. (Only lack of funds keeps me from going bananas ordering things.) Check your Whole Earth Catalog, Epilog, and past CoEvolution Quarterly's for other sources. Most mail-order places accept returns gracefully.

For faster service or for items awkward to ship, hit the nearest big city wholesale hardware and industrial supply houses. They usually will only sell wholesale, so you'll have to work out something. It can be very frustrating to have to pay 40 percent more for something than it is worth. Behind the wholesale curtain, we call it. A contractor's license helps.

As thing-makers, tool freaks and prototypers, Kathleen and I find ourselves custodians of about a ton of versatile hand tools. These have been used by us and friends over the years to help many projects and repairs get done. People keep asking us what tools to get, where to get them, and how to keep them from getting ripped off. Well . . . here goes.

Stand in Sears' tool department and it'll soon be obvious that you don't need one-of-each even if you have the money. Ask a craftsman what to buy, and you'll get as many answers

were personalizing their surroundings, often to the point that made the houses surprising places to someone, say, two inches taller or (as I was) two feet shorter or, simply, not them.

That is a large part of why I love to work wood. I have the illusion that I shape my life. When I eat at my table or sit on my bench or work at my desk the satisfaction belongs to me. Selecting the pieces from a stock of things

AMPACITY

Ampacity is the ability of an electric cord to carry current at a useful voltage. If you work with power tools more than a few feet from an outlet, it's an important consideration for you. As an extension cord increases in length, its resistance to current increases and the "line drop" in voltage can reduce power and can even damage your motor. A larger cross-sectional area of conductor will carry current with less line drop, so a longer cord must have a thicker conductor. Conductor size is stated in American Wire Gauge numbers: the lower the number, the thicker the wire. Ampacity varies with other factors, of course (notably wire alloy and temperature), but these guidelines may help match an extension cord to a motor.

Motor Rating			Cord Length	
horsepower	amps	watts	to 50'	50'–100'
1/4 or 1/3	7	875	18 ga.	18 ga.
1/2	10	1250	18 ga.	16 ga.
3/4	13	1625	16 ga.	14 ga.
1	15	1875	14 ga.	12 ga.

Matching cord covering to use and location is as important as matching length and motor rating to conductor gauge. Damp conditions or deterioration due to oil or heat can cut voltage at the motor if the insulation and covering are not appropriate. The National Electric Code lists cord types by letter designation and use. Some of the most common are:

	covering/insulation	use
S	rubber	extrahard use, damp
ST	thermoplastic	extrahard use, damp
SO	oil-resistant compound	damp, oily places
STO	oil-resistant thermoplastic	hard, damp, oily surface
SJ	rubber	junior hard usage, damp, lighter gauges
SJO	oil-resistant compound	hard use, oily, damp, lighter gauges
SJT	thermoplastic	hard use, lighter gauges
SJTO	oil-resistant thermoplastic	light gauge, hard use in damp, hot places
SP	rubber parallel cord	light use
SPT	plastic parallel cord	light use
PD	twisted cord, cotton or rayon	light use, dry places

could not affect me in the same way. The work and time, and the mistakes, that went into them have made them storage batteries of a supportive current and I draw from them more comfort than the rarest antiques could generate.

I try to make things that fit my needs and curry my comfort. In this way I am reaching back to that wellspring of craft. I satisfy my

as people you ask, for each has his own favorites and specialized needs. All agree on one thing though: BUY THE BEST YOU CAN. And the more a tool will be used, the better the quality should be. Tools used every day, especially electric tools, should be of commercial or production-line grade. You usually can't find these at hardware stores. Industrial supply houses are where to go. Take a friend who can buy wholesale. These tools will be expensive, so we'd better justify the cost.

For many, the best reason to go first class is that good tools are a real pleasure to use and handle. This helps make work less labor. The heavy-duty stuff looks brutal. It wasn't made to look good in the box, it was made to do the job and has been perfected over many years. The tough ones have their own kind of beauty that you'll see better as your viewpoint gets aligned with reality. Such tools, of course, last longer and are repairable when they finally do wear. They can take a lot more abuse, especially the inevitable overload. They can handle the bigger jobs and poor working conditions that

own need with my skill, and I channel it with my sense of beauty. My own. I have pride in my learning and the feel of tools that really work, and the sweet smell of cut wood. I am closer to my life.

I am an irascible man, partly because I wake each morning disgruntled at not being Leonardo da Vinci, who had the mind and the skill for everything and seemed to have the

WEAR GOGGLES WHEN USING CHISELS OR PUNCHES

would soon trash cheap versions. And after a few years in your hand, they often get to be old friends.

For tools that get used now and then, middle quality will do. By that I mean Sears' better grades and no lower. Really cheapo tools are of no use at all, can be dangerous, and often break the first time you use them. They are also discouraging to use, which might even cause a beginner to give up. Our only regrets have been not buying the best when we could have.

Tools that receive great strain, such as gear pullers, should be super top quality only. If you only need one every five years, rent it.

Okay, so what tools do you need? How do you start the stash? There are a few basic tools that everyone should have available: hammer, crosscut saw, adjustable wrench, pliers, screwdrivers (get a set), tape measure, hand drill and bits. Beyond these, you'd best gather tools as you need them. Auto work will require a rather complete set of wrenches and a whole boxful of special tools, some of which are for

time for it, too. I can't do all I want to do. I have not the skill, nor the courage, nor the patience, nor the time to do it all. Perhaps you, too, are less than perfect? We cannot all be cabinetmakers and we must admit that. Then what is left to us? Two alternatives.

There is an intoxicating practice known to your great-grandparents and practically extinct in our generation. It gave them great

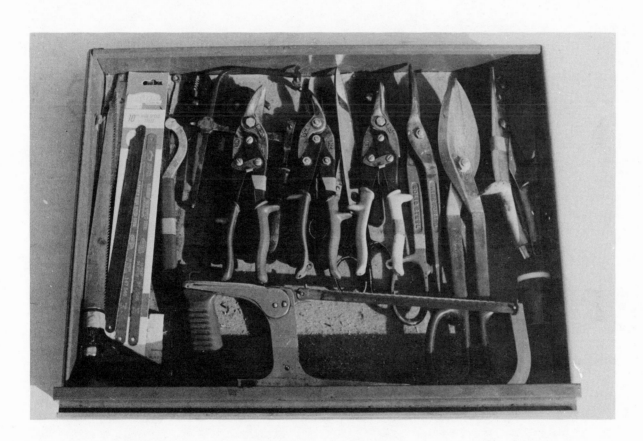

particular vehicles. Carpentry will require another whole group: planes, chisels, etc. Electrical and plumbing, still more. Our rule of thumb is that if we need to borrow a common tool more than once, we buy one.

Flea-markets are good places to look for expensive items like vises or anvils. Absolutely the best place to get a whole mess of tools at once is to keep alert for a widow selling off her deceased husband's retirement shop. Another place to look is auctions, but you'd better know what you're doing. You should shop around.

Recently in the Bay Area, we were quoted prices varying 50 percent on a tool we wanted! If you want to buy a bunch all at once (which makes sense these days of inflation — tools are a good savings account), some stores will make you a 20 percent deal. Even Sears can be dealt with, as the salespeople work on commission. They and other stores also have unadvertised freight-damaged goods hidden away. These can be good deals, as the damage is often merely cosmetic. You can give a salesman your name (and take his card) and

pleasure and it was as practical as a baked bean supper. They commissioned work. A table or a chair was needed, they went to a craftsman whose work they knew and had it made to suit them. If you have never commissioned a piece of work you will be delighted at the feeling of pride and accomplishment. It is the next best thing to making it yourself. In choosing a craftsperson, in defining your needs

have him call you when a certain tool is on sale or arrives damaged.

Whether in a big store or private sale, you should critically inspect each tool for condition. These days, many new tools by reputable (?) manufacturers are faulty. Used ones may be worn beyond repair. Anyway, be nitpicky about it; you'll be living with it in your hand. And beware of package deals claimed to be a great saving. The "complete mechanics tool set for $450.00" often includes tools you don't need, and may force you to take inferior items

that you would be better off picking up individually.

What do you do about that little voice that whispers, "Buy one, you might need it someday!" Well, it's *possible* you'll be needing them *all* someday, but Sears is only the tip of the iceberg. Have you ever seen a *real* hardware catalogue? Two thousand pages? On the other hand, it often does pay to get a set of tools that greatly increases your capability, such as a welding rig. Another way to go is for a group to buy a set of tools for working on one partic-

to him or her, in choosing woods and finishes and in reviewing sketches intelligently — being part of the design rather than the designer — you are sharpening your appreciation of the craft and taking part in the assessment and compromise that marks the difference between craft and machine work. The delight at having skilled hands shaping a cabinet or chair or desk to your personal

ular item, such as old Chevy 6 engines, and then everyone in the group who needs a vehicle gets one that has that engine and hence uses those tools and the consequent parts pile. (That's being done around here. There must be *dozens* of 1956 Chevy pickups and flatbeds within thirty miles.) Some groups get known for specialties: "the Butterfly Mountain people fix tractors." Those communities and families pool their resources and buy a set of expensive heavy-duty tools maybe for tractor repair. You have to be pretty mellow to make this work, especially if there is a high turnover of people. But this is a growing trend, and we think a good one. It leads to barter and lessens the need for duplicate sets of specialty equipment.

Our shop is known for its versatility. It's portable; everything fits a 4 × 5 U-Haul trailer. It's been set up in ten different places in five and a half years. The tools were chosen for quality and versatility. With versatility goes a handy ability to work in harmony with other tools, enhancing all. For example, with the drills and vises we have, we can drill a hole at any angle in just about anything. The combinations allow us easily to mass-produce parts like dome struts or Inkleloom frames. This gives a nice potential for making money as well as greatly easing tasks that might be as bad as working in Detroit. Versatility also means needing fewer tools, which means less money out, less space for storage, and fewer tools to keep track of.

For many people, the biggest problem with tools is keeping them together. That was our problem too for a while, especially at Pacific High School where there were always a number of young people who didn't yet see that tools are in a different category from other possessions. Our answer has been to take the time to try to give people a good feeling about tools being extensions of their own hands, and that tools are the means to getting good shelter and other desirable results. A French poet (whose name I regrettably can't remember) said, "Hammers spend a lot of time sleeping. . . ." We like to see the tools at work. We show people how to use the tools and encourage them to in turn show still others how. Having good tools in the hand, together with that tasty feeling that comes from teaching somebody else, gives the tool borrowers a respect for the whole bit.

We also have all the tools marked with a colored stripe. This not only reduces arguments on job sites where lots of people's tools are at work, but it makes it easy for people of good heart to return strays. We put out the word: "Bring a blue-striped tool to breakfast" and we round 'em up. We also ask that tools be brought back at sundown unless needed that night. There's a place to bring them back to. This is essential. A casual pigpen shop just can't keep its tools because there "isn't any there, there." As an experiment, we abandoned our collapsed old bureau toolbox and bought a (freight-damaged) Sears (the best for the money) rolling mechanics tool chest like you see in big auto shops. We segregated the tools by function and labeled the drawers. The result is that tools are easily looked over and selected and just as easily put to bed. To our

desires is so infectious that it may well encourage you to commission work from other crafts: weavers, tailors, potters, bookbinders, bootmakers, sailmakers, knifemakers, photographers, gardeners, and painters. Every craftsperson I know prefers commissions to the open-ended chore of turning out piecework; it is an exciting and very human way of doing business. The cost can be greater, but the

TOOL COMPARISONS

He is very careful, almost thoughtful. He plucks and peels, dropping leaves, making a good job of it. When it is fit for use he examines it again and then inserts it delicately, a third of its length in the opening. He withdraws it, covered with termites, and just as delicately, almost thoughtfully, he eats them. He is a toolmaker, a chimpanzee setting himself above the skills of all other predators and prey by adapting to needs with something other than his body or his long-memoried genes; he adapts with a prosthesis, an artificial body part, a tool.

We are toolmakers and tool users. Our prosthetic skill — skill at supplementing our proficient and versatile bodies with tools — is one of the signal advantages we hold over the bulk of the natural world. We began with peeled sticks and fire-hardened wooden points and chipped flint. We passed on to subtle bows and catapults, iron hammers and pod augers, winches, pole lathes, and the machined screw. How many hammer patterns have specific trades developed? How many axe patterns did a single trade generate? From a time of hand craftsmanship, when ingenuity had to outflank drudgery, how many tools are common today? How many tools are recognizable: half? a third? And what kind of tool is the electronic calculator or the home computer?

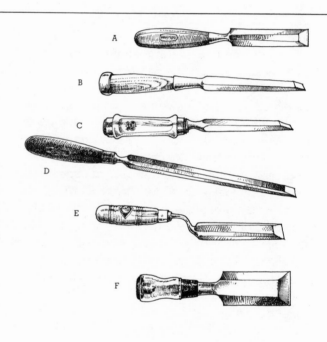

A. Bevel-edged cabinetmaking chisel
B. Socket-paring chisel
C. Mortising chisel
D. Long bevel-edged paring chisel
E. Cranked-neck patternmaker's paring chisel
F. Butt chisel

value is greater, too. When you engage a professional to work for you and with you, you bend his aesthetic and his problem-solving ability to your needs, to enrich your surroundings, and that is worth some premium. When the piece is finished, whether it is a tunic or a bed, it will be carefully made of first-rate material, overseen through every step of its making, and it will give more service and

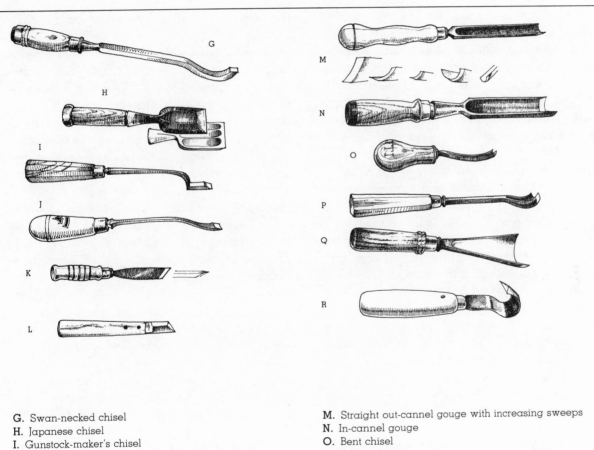

G. Swan-necked chisel
H. Japanese chisel
I. Gunstock-maker's chisel
J. Bent chisel
K. Skew chisel
L. Chip carving knife

M. Straight out-cannel gouge with increasing sweeps
N. In-cannel gouge
O. Bent chisel
P. Spoon gouge
Q. Fish-tail gouge
R. Swiss carving hook

stand more strain than any manufactured counterpart. That is worth a premium. If you take quality and labor into account you will be surprised how little the premium is. Taken by itself, that part of the price will afford you as much pleasure as any investment you could legally make.

If you commission work, your appreciation of woodcraft will grow. You will see grain and

great surprise, we found that this chest caused a drastic increase in the number of tools being used and a similar increase in action. We even found that we were using our own tools more! The neat storage made it easy to see who was missing, but people brought them back much more reliably than before anyway. The chest can be locked to control unannounced borrowing, which is always a disaster. The overall effect has been that under very poor risk conditions, both sociological and physical, we've only lost about fifty dollars' worth of tools in seven years! And this without having to get too heavy or "high school shoppish" about things. In case you wondered, we did try the tool-board-on-the-wall. It didn't work, and nobody we know that's tried it has made it work either, though it is nice to see all those tools hanging. It has not been necessary to sentence anyone to being tool-crib librarian either. We'll admit that it takes some time to develop tool consciousness in a crew, but it can be done, and peaceably. The tools spend a lot less time sleeping, too.

Making a deliberate effort to raise your own tool consciousness can result in some interesting new possibilities in your life. As with most mysterious-appearing phenomena, a bit of learning soon clears things up and you wonder what had been previously keeping you from doing your own repairs and thing-making. Sometimes all it takes is a different point of view. I've remarked that tools are extensions of your hands. No mystery there: a hammer is just a hard fist; a screwdriver, a tough fingernail. But hands usually operate according to instructions from head, so it can also be said

that *tools are an extension of your mind.* Looked at this way, the big (expensive) mechanic's cabinet with all the tools of similar function stored together with high visibility becomes even easier to justify. I find it is effective to store the tools by function rather than by name because this is the most useful way to think of the best tool when you are selecting. Hitters, grabbers, slashers, abraders — regardless of what they are called — are there in their places. You take your pick. Often, just looking at them will give you a better idea not only of what tool to use, but how to do the job or how to design the object. That's a big advantage of the neat toolbox. If the tools are "somewhere out on the back porch or maybe in the back seat of the VW" then your mind is deflected from creative thinking into a hunting mode, and the aggravation can easily cause you to lose your ability to get things done.

Easily accessible, functionally sorted tools also give you a ready familiarity with the tools you have. This has two effects. First, as you get to know your tools, you gain the easy fluid motions that go with using them. You're not afraid of them anymore, though respect is increased. This makes you able to work faster with less fatigue just as good form in sports often makes a big difference. It makes you safer too. Safety is also enhanced by having the tools where you can easily inspect them for condition, sharpness, and rust. We have found that safety is largely a matter of attitude. The closer tools are to being a working extension of your mind, the safer you'll be. Self-preservation.

The second effect is that you get to "know"

joints and turnings all around you and you will undoubtedly want to try your own hand at it. Once more, we are against the bald fact that not all of us can be cabinetmakers, and come to the second alternative: don't be a cabinetmaker.

The point is that you don't have to apprentice seven years as a cabinetmaker to be part of the satisfaction in working with wood.

all the tools you own without having consciously to think about it. This makes it simple to round up strays, of course, and it's easier to see where there are annoying gaps in your capabilities ("We don't have a lightweight mallet"). More importantly, you begin to think in terms of the tools you have. The eventual result is that you and your entire toolbox and shop become a big, complex tool with many possibilities. You begin to sense what you can do together. Buying tools with overall flexibility of purpose in mind keeps you from falling into the trap of building a one-function capability with accompanying tendency to fossilize conservatively your creativity. This tendency is strengthened as the value of the tools rises, which is the main reason society doesn't get fast response to its changing needs from large corporations that have sunk enormous capital into shops that make *that* only. Like fat cars. Once in that position, it's difficult to evoke at all, let alone without damage or drastic change of form.

So you begin to build your tool capability into the way you think about making things. As anyone who makes lots of stuff will tell you, the tools soon become sort of an automatic part of the design process. Beginners worry too much about skill and safety, rather as new drivers worry most about jerking the clutch when learning to drive a stick shift. It doesn't take long before more serious aspects take over, and the manipulation problems fade out. But tools can't become part of your design process if you don't know what is available and what the various tools do. In addition to buying tools that I find useful, I spend some

time reading catalogues so as to become familiar with tools that I can't afford or don't need at the time. Tool catalogues such as Silvo are rather like my cabinet in appearance so I find it painless to sort of automatically file the information away in the back room somewhere. Tool dictionaries, especially of older tools, are helpful too.

Some of you are saying about now, "Who wants to get into it that far anyway?" Friends, there are advantages. Obviously, making or repairing things yourself can save you money and time. Well, maybe it isn't so obvious. Example: next time your car breaks down, find out how many hours it will likely take to fix it. You don't have the time, right? Okay, how many hours will you have to work at some job so you can pay that mechanic? For many of you, the hours you have to work to pay the mechanic will be more than the job would take if you did it yourself. Moreover, you don't have to pay yourself, and the job can be done to your standards and at your convenience. If you don't have the skills or the tools, that's what we're talking about! Doing it yourself can free you from certain dependencies that you may find smothering. What if the twenty-dollar-an-hour plumber can't come until next Friday? Repairing pipes is relatively easy. Once you learn how, you not only avoid being at someone's mercy, you have a skill that can help friends or make money for you. How-to books are tools, in case you haven't guessed.

Another advantage of having some tools that you know how to use is that as you get an easy, facile familiarity using them, you begin to get a better feel of the ergonomics of other

If this book, this bouillabaisse of woodwork, is to be a success, it must make one point distinct: there are more levels and areas in woodworking than kinds of wood, and each has its own qualities and cautions. Each is honorable and enjoyable of itself. Fine cabinetry is not the pinnacle of woodworking, it is only one of the things a woodworker can do.

There is a hardware store in Columbus, Ohio, called Columbus Hardware. It is a high-ceilinged storefront downtown, but when you walk in out of the slicing December wind that scours Columbus, the high simple space is immediately comforting. It is chiefly the wood that effects that cozy feeling, the worn wood floor, the floor-to-ceiling bank of wooden drawers mounted by ladders that run the depth of the building on tracks, and the repetitive wooden tool handles in racks. It must be the wood because the store is not overly warm; the men behind the long wooden counter wear sweaters. They are serious, pleasant men, each with a shirt protector stuck with a ball-point pen and one or two pencil stubs, each wearing on his belt or carrying in his wool pants a tape measure or folding rule. They are usually busy. December, though, is a good time to walk in; Columbus Hardware sends out no Christmas circulars, no phone message pads for customers (I do not believe they advertise) but they do have a Christmas offering: near the rear cash register is a carton of eating apples from an orchard near Cleveland, perfect scarlet Macs cradled in eggboard separators and square yellow tissues. Beside them is a carton of rolled calendars, the large, shop-wall size, with a print of a country scene, "Cidering" or suchlike, and under it COLUMBUS HARDWARE. (This may be advertising.) You can take an apple, examine the calendar, and look around while you reduce the Mac to a thin core. There are table saws, radial arm saws, drill presses, and mounted bench tools standing on the floor with their cords coiled. There are outdoor tools: a dozen patterns of shovels, hoes, grubs, peavies, Johnson bars, scythes, hayforks, and some implements you won't recognize. This is no difficulty; someone — not necessarily a clerk — will be glad to name it and explain it. The atmosphere here is like a men's club: relaxed, you need not know someone to speak with them. When a clerk is free he can help you, not always by selling you something. He may listen to you describe your project, nodding, and mount one of the ladders, returning with fittings that may help. You can discuss it with him. He will know. He can compare power tools, discourse on concrete pouring, or advise on porch building. He may call in an expert in the form of another clerk. The prices are fair. The inventory is large. In the back, near the alley doors, glass is cut. In the basement are cordage, bulk nails, parts, and stock. There is a glass case near the street door: practical knives, reels, micrometers, small tools, tape measures, specialized tools. It is difficult to leave without stopping there.

There are not too many places like Columbus Hardware. It is a reasonable place, a source, a reference, a workable anachronism. My local hardware store is mostly Formica and white pegboard. Its tools and screws and, damn, even its nails are in blister pacs. It won't sell me T-nuts or hog rings or timber jacks or a set of trammels.

Are such things necessary? I don't know. I do believe that one of the walls of plastic society is buckling: people are still insisting that they can use their hands. People are finding it (increasingly) necessary to prove their own usefulness and skill in physical statements. Men and women and young people want to make part of their world.

Tool is a noun and a verb, and any human time is caught up in the motion of the verb, never complete but always on the way to mastery. Tools are beautiful: they have scale and direction and economy; grips invite your hands, edges point at the work, and any unnecessary or decorative part is — for better or worse — instantly recognizable as something outside the skeleton of function.

It is rare now to inherit tools, there are too few craftsman-fathers to go around. We have the catalogues, though — our collective, expensive grandfathers. Catalogue-cruising is a pursuit of its own, an exercise in cerebral cabinetry, and many mail-order joiners who couldn't drive a tenpenny nail without bloodletting can identify specialty planes that would mystify damn good jackleg carpenters.

Perhaps it's just as well that we don't inherit our fathers' tools. Fifteen years ago a garden-variety consumer would be hard pressed to buy a plane or cabinet chisel of good quality. It was the offslope past, the do-it-yourself peak of the late fifties when "home handyman" was a title that

The rough carpenter works with dressed lumber of nominal sizes, framing out a building in a way that brings up walls and a roof and figures in the space with rooms at a speed that would confound a cabinetmaker. The rough carpenter's special skills are large: scaffolding, squaring up, bracing. His tolerances are loose, a sixteenth to an eighth of an inch

had a snappy madras and Weejuns sound to it. Whacking up an ironing board holder or a shoeshine kit was good fun and about as cerebral as shuffleboard or canasta. We are a new kind of craftsperson, and it is a time that looks at the inner game of joinery or carpentry with perhaps more interest than at the finished product. It is also a time when running down the street involves its own rationale, wardrobe, stopwatch, and mantra, and with the blessings of availability we must deal with the dangers of mystique. There is an implied obligation in all the catalogues: we view them as a tool chest and feel obliged to have the "proper" tools before we can do proper work.

You must puncture the mystique dragon in your own way, but for me the dragon shrinks a little when I think of a Japanese carpenter marching to a job with a small, a very small, chest of tools. I also think of Fred Daley, a small old man and a good house carpenter, walking away from a finished house with one open box of tools, not a large box. Begin with this: with a dozen hand tools of good quality, and with good wood, and with concentration, you can do work of extraordinary beauty. The skill lives in your hands and your will and is extended by the prosthesis of tools only so far. Most specialized tools are conveniences and refinements of simpler devices, and are in all probability superfluous. Shrink the dragon in your own way; the beauty of tools is at odds with the pocketbook, with practicality, and even with the learning of simpler skills, but it is an enjoyable narcotic.

With due caution, then, you can order and wade into all the tool catalogues and what will you find? Which is best? Which offers the best value? Which is most complete? The answer is not as simple as a stack of numbers but we can address the simplest question first, that of prices: a survey of identical brand-name tools priced in five or six catalogues will show that prices do vary, inconsistently, and that you must shop for tools among all the suppliers, ordering here for mallets and there for nailsets. One generalization available is this: some big-volume tool supply houses, like U.S. General and Silvo, carry more abstruse

tools than might seem likely, and you may make considerable savings by ordering what they carry over woodworking specialty houses like Woodcraft, Garret-Wade, Leichtung, Princeton, and their like. Silvo Hardware Company, for instance, carries the standard Stanley planes but also stocks the Record planes and even one make of wood-bodied planes.

Another generalization involves power tools. Large supply houses, dealing in volume, can give a better price. Here, as with power tools, you must shop, and that is the long and short of it.

There are times when the tool mystique grows into actual mysticism, times when shrewdest shopping can't help you and you must go on intuition. This is in the realm of material quality and preference. A twenty-six-inch, ten-point handsaw is not a definite item: it can be a Disston, a Nicholson, a Nonpareil, a Saandvik, a Pax. How can anyone tell you which is best? Nonpareil sawyers look on Pax sawyers with pity and both look on Disston sawyers with scorn, yet all three may get through the board with the same speed and accuracy. Steel differs, temper varies. No one can tell you which chisels to buy. Barely polite controversy goes on as to which of four or five long-defunct chisel makers were best, chisels pass from hand to hand with large sums of money, counterfeiters appear. Carving gauges present a more confusing situation since handles, shapes, and lengths are as unstandard as steel, temper, and edges. Japanese tools are, for the most part, handmade; their quality and reputation differs in the same way that samurai swords had the legends of their makers and their exploits. A Woodline saw may be four times the cost of a Toshiro of the same pattern . . . and it may be worth it.

This can be said about quality: it takes a master to bring good work out of a poor tool. If you are not a master, buy the best tool you can afford commensurate with its use. We are not the only beneficiaries of the woodworking renaissance: our children and our grandchildren can inherit tools from us, if we choose and use them wisely.

at times. His work is hidden behind walls but it must be strong, and he must close and roof over as quickly as he can, because weather is one of his variables. To watch good framers at work is to see a dance with few small steps, the rhythm beaten out with a twenty-ounce claw hammer. They banter and kid each other about elephant feet (the face mark around an

techno-hardware that you use or make. (Ergonomics is the man-machine interface: how the steering wheel feels in the hand and how it tells you what's happening to the wheels; the wrist-breaking poor feel of eggbeaters; the built-into-you feel of a good rifle.) Poor ergonomics is one of the main reasons behind the recent public disenchantment with technology. Things are made with the convenience of the machine in mind instead of the human user. The result is hardware that is hard to hold, too cold or too hot, difficult to repair, easy to lose or lose control of, easily broken, etc., etc., etc. The machine is in control of you instead of the other way around. You can do better, yes? Most highly evolved good-quality tools are ergonomically good. So without having to take a course in the subject, you can gain an informed feel of what is satisfying. As with most problems brought to us by technology, ergonomic problems are often best solved not with more technology but with clear thought and a better-informed intuition.

An informed tool intuition works best if it's augmented by an informed materials and processes intuition. For instance, if you don't know anything about foundry work (casting), it's unlikely that you will come up with ideas that require it. Often, this ignorance (ignoreance) is easily remedied. A bit of inquiry may well show that what you had considered a black art is actually not one at all. Bronze and aluminium castings, for instance, are made every day in high school art departments by unskilled students using scrap metals from auto wrecking yards. Anyway, things take a

form dictated by the possibilities inherent in the material to be used and the tools that can shape it, and the ideas in the head of the worker. It follows that the more you read and snoop around and experiment and practice, the easier it will all come and the more independent you can be. Freedom rising.

The ultimate is to make your own tools. Tools fitted intimately to you by you. What could be niftier? Blacksmiths are really into that. A good example is found in the books by Alexander Weygers. But tools need not be limited to the shop. How about making your own personal canoe paddle? Or your own left-handed kitchen equipment? You can modify existing tools too. For instance, when we needed to make seven-inch-diameter pistons for a small production run of giant raft-inflating hand pumps, we reversed the bit in a hole cutter so it made discs instead. The pistons were then easily and accurately cut from heavy plywood with a great saving in time and material compared to turning them on a lathe that we would have had to borrow. Making the big pump's leather "piston ring" seals proved to be easy after we spent some time talking to craftsmen in a sandal shop. With their advice, we soaked heavy leather discs in mink oil and then pressed them into the desired shape with a matched male and female die rammed by our vise. The dies were made on a band saw modified with a simple homemade attachment that enabled us to cut beveled round holes with good accuracy. That attachment was also used to make the next batch of pistons, as it proved faster than the disc-maker on the drill press.

overdriven nail), their tools are hung around their waists in leather belts, their circular saws are loud and the blades ring after the motor stops, they keep an eye on the clouds. Their work is strong and beautiful.

The cabinetmaker works quietly with many pauses. He cuts and planes rough-sided planks of special woods to size only after long

Tools making tools making tools.

I can hear some of you saying, "Small production run! Yuk!" Unless you are an artist, and maybe even then, you will sooner or later need a bunch of things all alike. Even with only the most basic tools, you can mass-produce things. The precision and complexity of the produced parts is somewhat dependent upon the adaptability and quality of your tool bank, another reason to gather good stuff intelligently. Large-scale mass production tends to enslave both the workers and the customers. The workers are used as if they were machines. The huge capital outlay for the factory means that there must be a huge and relatively steady demand. This in turn means heavy manipulation of public "desires" and almost always involves politics and coercion.

Small-scale production, though, can mean a great reduction in drudgery as well as interesting possibilities in barter. By means of jigs and other simple fixtures that you can figure out yourself, you can be freed from having to measure each part. Hold the piece of wood against the jig and hit it with the drill and all the holes will be in exactly the same place in each part.

We've mass-produced thousands of dome struts and parts for conventional construction. We've also produced simple looms and the aforementioned pumps (which we bartered for a fleet of rafts and got into the whitewater river-running business), solar collector parts, signs (during an anti-freeway fight), toys, boxes, electronic parts, concrete forms, fence rails, adobe blocks, tents, shelving, lighting systems, and model parts, to name but a few. The ability to get on a small production run frees you from dependence on larger, less efficient manufacturers, their prices, specifications and schedules. It can be rather fun, also, if it doesn't go on for too long. "Shop Yoga," we sometimes call it.

And it can be done without sophisticated expensive equipment if you take the time to think it all out first. The thinking is the most powerful part. You can actually *change* some things out there! Maybe not in a big way, but certainly at a scale that you can understand. You will find that as you work, your understanding of technology will increase a bit, and your fears based on ignorance will decrease. (Your fears based on newfound understanding might well increase, but that's another paper.) In a modest way, you can combat the "machines taking over" by having better control of the technology you live with. It's a good feeling. And it's free.

Jay Baldwin is a genetic technocrat, a man designed by nature to use tools: a compact, powerful torso with ground-sure legs, thick canoeist's arms, and enormous unquiet hands under a great head. The manner is almost forbidding at first but the stories, opinions, and facts that come tumbling at you beguile and, often, delight. He is a designer, a student and colleague of Buckminster Fuller, an inventor, an ardent advocate and influential worker for appropriate technology and energy use, and technical editor for the Whole Earth Catalogs *and for* CoEvolution Quarterly *(in which this article originally appeared). Young people like him and, of course, he plays the musical saw.*

deliberation, after more than a nodding acquaintance with the figure and grain of his piece. He has the radio tuned to a quiet music station. His special skills are finish, angles, fit, and design. His special tools are paring chisels, many planes, gauges, marking knives, and he would sooner have you borrow his

First Aid in the Toolbox

by Dr. Matthew Finn

Anyone who works with tools or machinery is, at some time or other, going to be impressed by the terrifying fragility of human tissue, just as any congress of experienced carpenters will reveal that few of its members have all their digits. The workshop is filled with accidents waiting to happen and, of course, with the advent of power tools the injuries resulting from these accidents have become more impressive.

Before I discuss specific wounds, some general observations are in order. With most workshop accidents, bleeding in greater or lesser degree will be the initial concern. If the injured part is in the hand or the arm, simple elevation of the wound above the level of the heart will suffice to control the bleeding in many cases. With more profuse bleeding, direct pressure on the bleeding wound may be necessary. Here it is important to point out that the common mistake of using a bulky towel or bandage over the bleeding wound makes the application of adequate localized pressure virtually impossible; a small concentrated bandage works better. Tourniquets should almost never be used short of a catastrophic guillotining. They are usually unnecessary, almost always improperly applied (so that they may actually increase the bleeding), and can even cause severe nerve damage when used by the unwary. When applying pressure one should try to resist the temptation to peek at the wound every few seconds. The body's clotting mechanism takes several minutes to work and sustained pressure for a good ten minutes by the clock is indicated before taking a look.

The prevention of infection is the next concern. There appears to be an overwhelming need to apply some magical glop to the wound. Let an injury occur to the hand and directly the wound goes into the mouth where it is presumably sucked clean. The human mouth, of course, is a veritable sink of iniquity, teeming with harmful bacteria; and this practice, doubtless done in imitation of one's dog, is utter folly. Rather let the dog itself lick the wound. But better, the generous application of soap and water, remembering that the goal is the mechanical washing-out of bacteria and debris. Unguents, ointments, salves, and balms, whatever their reputation with the local shaman, should be avoided. They obscure the wound, thus preventing subsequent inspection, and may impede adequate drainage

toothbrush. His tolerances are tiny, less than a sixty-fourth. He is patient. He shaves away a translucent curl of walnut and tries for a fit. He reclamps the piece in his vise and shaves off a thinner curl. His work is strong and beautiful.

— always a desideratum in contaminated wounds. A dry sterile bandage is all that is necessary.

If the wound is more than a puncture, the question arises as to whether or not it will require stitches. This decision is probably best left to your physician. If a wound is to be sutured, the sooner it is attended to the better. Any time-lag increases the hazard of further contamination and infection.

Wounds of the Hand and Forearm

The human hand is an intricate marvel; but all its delicate structures lie so close to the surface that they are terribly vulnerable to what might at first seem even trivial injury. A slipped screwdriver driven into the palm can sever a nerve or tendon and in an instant convert that hand from a fine instrument into an awkward spatula. A nerve lying a few millimeters beneath the pathetically thin skin of the wrist may be cut with absentminded ease and so end a promising career. Any wound of the hand, except those that are obviously superficial, must be viewed as potentially serious and should be examined by a surgeon.

Power tool amputations are seen with increasing frequency. Although newer microsurgical techniques are occasionally used to reattach amputated digits, this is usually not feasible. Nevertheless, the amputated part, wrapped in a clean or sterile cloth, should be brought to the hospital with the patient. Obvious flailing of fractured bones can be pre-

vented with splinting. Foreign bodies, most commonly wood splinters, are frequent invaders of the hand's soft tissue. These will almost invariably become infected if not removed. Every seasoned woodworker knows that if he waits awhile the splinter is often floated out on a tide of pus. Warm wet compresses may speed up this process.

Eye Injuries

Next in order of frequency are injuries to the eyes. Safety glasses, alas, are viewed by woodworkers with the same disdain that hockey players reserve for helmets. Penetrating wounds of the eye quite clearly call for professional attention. More insidious are foreign bodies on the surface of the eye. These may become, if not dealt with properly, ulcerations of the cornea with delayed scarring and visual loss. If, after removing a mote from the eye, discomfort persists, a physician should examined the cornea for signs of ulcer action. Chemicals in the eye should be washed out with copious amounts of water.

Burns

Most burns received will be partial-thickness burns — so-called first-degree and second-degree burns. These are characterized by pain and redness and, with the deeper burns, blistering of the skin. Plunging the burned skin into cold water will help relieve the pain. If blistering is present or immanent a dry sterile dressing will protect against bacterial contamination and will reduce the discomfort. Here again, ointments are to be avoided, whatever your grandmother may advise in the way of burn butter. Full-thickness burns, not uncommon in electrical burns occurring in the shop, should be dressed with a dry sterile bandage and a physician should be consulted.

Cardiac Arrest

Cardiopulmonary resuscitation (CPR) is developing into such a universal skill that it is becoming very difficult to die suddenly anymore. The presence of power tools and the attendant risks of electric shock make it important for the woodworker to understand something about CPR. Inadequately grounded appliances are a common cause of death by electrocution. CPR is a technique designed to restore, artificially, the circulation of the blood and respiration. If a person, struck down by electric shock, is not immediately revived by three or four sharp blows to the chest over the sternum or breastbone, then CPR should be begun. The victim should be placed flat on his back on a firm surface. The chin should then be lifted and the head tilted backward in order to open the airway. If the victim is not breathing, mouth-to-mouth respiration is begun immediately. Start by giving four quick breaths; then check for the presence of a pulse in the neck. If there is no pulse, start cadiac compression in order to restore circulation. Using the heel of the hand, reinforced with the other hand, fifteen compressions of the sternum are made at the rate of eighty per minute. Between each cycle of fifteen compressions, move back to the head and deliver two quick full breaths. This should be continued until the victim responds or help arrives.

Dr. Matthew Finn is a respected surgeon, a canny sailor, a poet, and an enthusiastic cynic. His dog, Sundance, worships him and many of the rest of us see him in a favorable light.

CATALOGUE ACCESS

Brookstone
127 Vose Farm Road
Peterborough, New Hampshire   03458

A slim, dense catalogue which arrives with alarming frequency, filled with many gadgets, many notions, and many tools of fairly good quality. The Brookstone catalogue has a style of its own, tending toward headlines like "old world craftsmanship" and "pride and joy" and "ultimate." Somewhat overpriced but a resource for odd items normally unavailable. Excellent, fast service, shiny satisfaction.

Conover Woodcraft Specialties, Inc.
18125 Madison Road
Parkman, Ohio   44080

Not a complete array of woodworking tools but all of a consistently high quality, including their handmade specialty planes, lathes, and screw boxes for cutting threads in wood. See page 114 for Ernie Conover's article on toolmaking.

Frog Tool Company, Ltd.   (50¢)
541 N. Franklin Street
Chicago, Illinois   60610

A broad line of woodworking tools with good selection in planes, carving gouges, log-building tools, and books.

Garrett-Wade Company, Inc.   ($1.00)
302 Fifth Avenue
New York, New York   10001

Some special mention must be made of this catalogue. It is the stuff dreams are made of, the very green eye of the mystique dragon, because the design and most especially the photography are beautiful. Each page is a composition wherein each woodworking tool gleams like a jewel. A page of simple files and rasps is an image of sharp iron you will not soon forget. Depth in choice and special depth to the selection of workbenches. A dangerously appealing catalogue.

Leichtung   ($1.00)
701 Beta Drive
Cleveland, Ohio   44143

Fine woodworking tools in an attractive catalogue with good selection. Specializes in the Lervad workbenches.

Sears
Call local Sears for catalogue

Tools for general carpentry and power tools, notable for their guarantee: if one of their Craftsman tools breaks or malfunctions you take it back and you get another one. No questions. This also makes for low prices on reconditioned power tools at a regional Sears center. Check on it.

Silvo Hardware Company   ($1.00)
107-109 Walnut Street
Philadelphia, Pennsylvania   19106

The best of the large supply houses available to non-wholesale consumers. A well-organized catalogue with an astounding range, including a good number and selection of hand woodworking tools. Strong in power tools. A necessary catalogue.

Tashiro Hardware Company
109 Prefontaine Place
Seattle, Washington   98104

A newsletter of inexpensive Japanese woodworking tools. A fair selection and probably a good way to try the tools.

U.S. General   ($1.00)
100 General Place
Jericho, New York   11753

A big outfit with a big, haphazard catalogue. Not a very good range of fine woodworking tools but best prices on power tools.

Woodcraft   ($1.00)
313 Montvale Avenue
Woburn, Massachusetts   01801

Woodcraft is greatly responsible for fine woodworking rebirth. They made really fine tools available. Even if loyalty counted for nothing, Woodcraft still has the broadest, deepest selection of hand tools now available. Another necessary catalogue.

Woodline/The Japan Woodworker   ($1.00)
1004 Central Avenue
Alameda, California   94501

A well-designed catalogue of exquisite — and expensive — Japanese tools, with a sprinkling of Occidentals. Sculpture that works.

CATALOGUE COMPARISON PRICES

* ITEM	Woodcraft	Mittermeir	Garret-Wade	Frog	Leichtung	Pringeton	Silvo	U.S. General	Sears Comparable Tools
8mm Skew Chisel	$8.80	$5.85	$5.60 (3/8")	$5.50	$6.50 (3/8")			$5.40	$2.99
8H 10mm Gouge	9.50	7.35	6.10	6.55	6.90				
5 Spoon 8mm	8.80	8.70	6.90	6.70	7.00				
Beech Mallet, 5" head	8.60		6.90	6.00		$11.00 (4")	$5.75		
Chip Carving Knife	3.85	4.80	3.60	3.35					4.89
Cabinet Rasp 10" 2nd cut	9.20	7.50	9.90	7.30			6.15		6.79
Norton Multi-Stone 2 grades crystolon, fine India	76.90		79.50			79.95			
Combination India Stone	8.60		7.90	8.00		7.95	7.19	4.95	6.99
Sliding Bevel Marple 5–9"	8.50 (9")		9.90	7.60 (8")	6.50 (8")	11.99 (8")	6.89	11.80 (Disston)	14.59
Try Square Marples 9"	10.35 (6")		8.90	10.90	7.50 (8")	9.95 (6")	8.98	8.50	
								9.90	8.79 (8" open)
Face Shield	9.20		7.90	9.30		8.50	4.75		
Hand Saw 26" 10 pt. Crosscut	27.50 (Nonpareil)		23.90 (Pax 8 pt.)	18.15 (Pax)		19.50	11.29 (Nicholson)	6.50	6.79
Back Saw 12" 15 pt.	29.70 (Nonpareil)		21.90 (Pax)	19.50	22.95 (Nonpareil)	13.50	9.75 (Disston)		21.39
Handscrew Jorgenson 10" × 6"	19.70		13.90	15.50	14.00	8.50	8.89 (14", 13 pts.)	4.70	5.39
Pipe Clamp Pixt. Pony	9.95			9.10			5.59	6.60	
Jack Plane 14" × 2" Record	33.95		29.50	35.85			27.29	13.50	
Pry Bar	7.80						4.55		
Bench Chisels Marples Blue Chip 1"	6.55		5.90			7.25	4.85		
Butt Chisels Stanley 1½"	9.70						6.39		
Forstner Drill Bit 1"	16.25	15.00		14.47	13.95		12.05		4.77 (1¼")
Rockwell Contractor's Saw, 12" w/o motor							499.99	439.95	
Rockwell Var. Speed Drill 3/8" × 2 frame							24.99	21.25	
Skil 7¼" Builder's Saw							125.00	106.25	
Rockwell 7" Dis Sander							149.99	121.50	
Rockwell Speedblock	66.95						64.99	47.80	
Rockwell 1 HP Router	78.20						75.00		

* Lowest price in italics, excluding Sears

The instrument maker works to old patterns. He bends wood to improbable shapes and thins it by degrees. No one uses better wood. He listens to it. His special skill is to make, out of dissimilar woods and shapes, a whole. His special tools are jigs and clamps and small cutting edges. He uses glue with a mind to future repairs. His finishes are carefully controlled. The first time he tunes up a

Flute Making

by John Ingalls

At first sight an article on flute making might seem esoteric to even the most curious woodworker, but there are many areas of overlap in woodworking. I remember listening attentively while a worker from the Martin Guitar Company described using a soldering gun to break the glue joint and reset the neck of a guitar. I knew that information like this gets stored in the memory bank of the brain and jumps out at the right time (you hope). This article may help you in a number of ways.

Material

The selection of wood for making flutes presents some special problems. As one can see from the diagram, the bore is moist and the outside is dry. This moisture differential produces stresses, causing the outside to check. As a result the flute maker is looking for wood with the following characteristics:

1. A low hygroscopic expansion rate
2. An oily imperviousness to water
3. High-density material that will produce an attractive finish
4. Nontoxic wood

Several years ago I caused the early retirement of a flute player. He loved the pao ferro head joint I built for him and chose to ignore the early symptoms of an allergic reaction. Months after putting the flute down he still had a puffed lip. He was unable to finish his senior recital at the Boston Conservatory.

Boxwood is a lightweight wood having a beautiful tan color. At one time it was common in Europe but a variation is now being taken from Thailand. Its high hygroscopic expansion rate is offset by its ability to expand without damage. Cornelius Ward, a nineteenth-century London flute maker, said that boxwood was more fitted for construction of a hygrometer than of a wind instrument. The fluthier Rockstro regularly immersed his entire flute in a tub of water to keep the tuning consistent while playing it. According to my measurements, boxwood even expands parallel to the grain; this problem makes it a poor choice for rulers, which is its most common application today.

African blackwood, a dense black wood, is highly appealing for flutes. I personally have found it unforgiving to work with. It has a tendency to crack. The finished flute is less sure of having good tone.

new guitar or violin he is quiet and apprehensive. The instrument must rigidly endure the sum of the strings' tensions, almost a ton. If it does, it is because his work is strong and beautiful.

Which has more honor? Which is club champion? Ridiculous. To each skill there is a unique satisfaction that enriches its worker. All workers are learners, and to learn is the

176

A guitar by Jeffrey Elliott with delicate carving around the sound hole

measure of progress. You can put up a wall or fix a faucet; there are good books to teach you how. You can make a chair for your terrace or a cold frame for your garden; there are books. You can build a dulcimer or a Shaker bench; there are kits and plans, you can learn from them. You can touch your own life, contribute to your own comfort, and become better by degrees and by mistakes. You will make mis-

Expansion crack in a wet bore

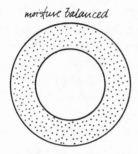
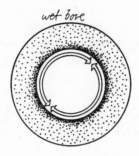
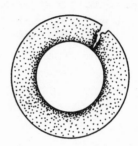

moisture balanced   *wet bore*

Pao ferro or *Brazilian ironwood* looks like walnut but is much more dense. It is easy to work with, produces uniformly good sound, and has never split on me.

Coca bola has a beautiful red color, is fairly easy to work, and has a mild tendency to split. This, too, produces an allergic reaction in certain people.

Pearwood is light in density, has an attractive light brown color, but is very porous. It will soak up oil indefinitely and I have been unable to achieve a tolerable sound without varnishing the bore.

Lignum vitae is unstable and difficult to keep from splitting while drying. Once bored, the wood becomes more stable. My one experience with a lignum vitae head joint produced yet another allergic reaction in the test pilot and I gave up using the wood. As a material it is similar to plastic and seems to produce a dead sound.

Plastic. In his treatise (1890) Rockstro pronounced ebonite the ideal material for making flutes. According to his writing it excelled all woods in durability, brilliance, and beauty. He continued to say that its ease of playing and

power made it especially suited for female flute players.

Several years ago a collector showed me a flawless flute by the maker Rudall & Carte. I will never forget the horror-stricken look on his face when I pointed out that the instrument was ebonite. I tried to console him by quoting Rockstro but it was no use.

Preparation of Wood

The aging of wood has several functions. During the aging process the wood contracts as it loses moisture. In some applications this shrinkage is not a problem, but an unaged flute will become out-of-round and all the fittings will fall off. The wet wood will not assume a good finish. The aging process tends to weed out the wood that is prone to splitting, and the advantage is that it splits long before it ever becomes a flute. I have talked with a flute maker who scrapes all the wax from a billet of wood to try to make it split. If it survives, it is a good piece of wood.

Several years ago I made a trip to Costa

takes, but they are marks of learning. You learn nothing from doing it right; only mistakes teach you, and perfection is boring. If you and James Krenov, that wonderfully skilled cabinetmaker, began at nine in the morning and worked until teatime, both of you would leave your shops with a day's satisfac-

Rica to gather coca bola logs for my work. The prime piece was a log ten feet long by twelve inches in diameter. For years it stood in the center of a corral where it had been used to tie cattle while branding. We chain sawed it into 36-inch lengths and used a band saw to cut it into 1½ × 1½–inch billets. I sealed the ends of the pieces and figured I was done with them for at least a year. The Costa Rican officials did not agree with me. Many countries have laws forbidding the export of certain prized woods. We argued and tried bribes, but to no avail. The wood had to be worked in Costa Rica before export. We made a compromise and agreed to have all the billets turned to a smooth finish. We took the pieces to a wood turner but it was no use. Have you ever tried to turn a warped piece of wet wood thirty-six inches long? We returned to the customs official and he agreed to let us out of the country only after the pieces had all been planed and sanded to a smooth finish.

I returned to the United States with four hundred pounds of lovely tropical wood specimens. Long pieces are a joy to work with because they can all be made to match in the finished product. Flute sections are so short that warped pieces can make a fine instrument. Since coca bola is toxic to some people, I found that the market for flutes made of this material is limited. Much of the load was sold for one dollar an inch in the form of nightsticks for the Somerville police.*

* Concerning the flute maker's shop: flute making is not a particularly lucrative profession. The shop should be centrally located in the poor section of a big city, in the slums. The flute maker is, however, guaranteed excellent police protection.

Ideally, the flute maker would like to age the billets for a year before doing further work. The waiting time can be drastically reduced by boring a hole in the center. Center drilled, reamed, and turned to slightly over finish dimensions, raw wood can be turned to a flute in less than six weeks.

Many of the old makers (before the band saw) split their wood. This process has the advantage of insuring that the grain is perfectly aligned with the finished product. This process is very wasteful of wood. The flute does not need to resonate like a violin, where fiber strength is important, and boxwood just does not split like Sitka spruce.

Boring

The best method of boring depends on the material and equipment available. Square billets up to twelve inches can be held in a four-jawed chuck and brought to roundness using a live center to hold the opposite end. Splits or irregular pieces are best brought to roundness between centers and then bored in a three-jawed chuck.

Shallow holes up to about twelve inches can be bored using jobber's drills. Deep holes are best accomplished using a gun drill. The hole is first started with a regular bit. The gun drill consists of a pipe with a bit attached to the end. The bit has a hole down the center through which air is injected. The air carries away chips and also cools the bit. This method of drilling insures a perfect hole with a mirror-

tion, frustration, and learning. There is a kind of equality, there, though you might have the edge on Krenov. You would have learned more, and you would be working on your own needs.

Mandrels: professional and occasional

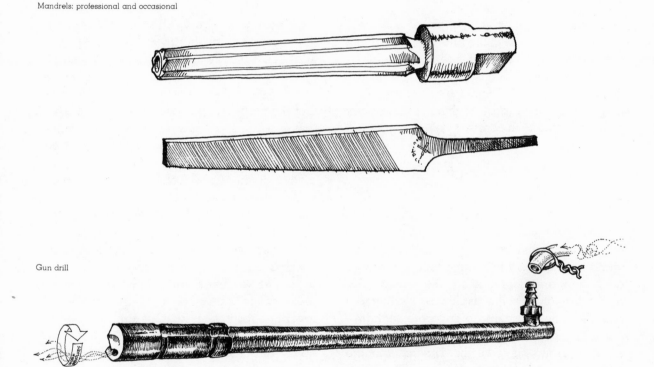

Gun drill

like finish. A variation of the gun drill known as the shell auger works on the same principle but without the injection of air. The twist drill relies on the precision of the point to guide the drill accurately. The gun drill point is off-center and thus plays no part in guiding the bit; it seeks the center as the path of least resistance through the work. If this article leaves one with any understanding of the gun drill it is much to the credit of the illustrator. I spent hours studying diagrams without a clue to its workings.

Reaming

Having digested all the material above, the reader has mastered all the skills of a Renaissance flute maker. In about 1680 the tone, ease of playing, and intonation of the flute were enhanced by tapering the bore. This tapering is best accomplished using a reamer. Tool steel reamers are cut to a taper on a metal lathe; fluting is then cut into the tapers on a milling machine. In a well-equipped machine shop the tool work involved in making a custom

A long time ago we began with wood because it was strong and light and workable. It has been one of the most important materials we've had. Our debt to wood is beyond reckoning and our feelings for it are deep. Professor Perry Borchers, architectural historian and acoustics consultant, opines that wood may have had an effect on the very sound of our

reamer takes at least four hours. The reamer is the heart and soul of the flute maker; without it he is nothing more than a repairman.

To the maker of one or two instruments the cutting of a tapered reamer is out of the question. I would suggest making reamers by grinding a file to the correct taper.

Turning

An essential to turning flutes is to be able to remove the piece from the lathe and replace it accurately and easily. For my purposes I found it satisfactory to use the tapered reamer as a mandrel for holding the piece. The method involves some risk to the reamer but cuts the time of making a separate mandrel.

To the expert turner of wood the outside of a flute presents no problems. The inside of the socket where two pieces join may be difficult. The best solution is to chuck the cylindrical piece in a large metal lathe and cut the mortise with a boring bar. If no such bar is available, the best solution is to chuck one end in a small lathe and hold the free end in a steady-rest.

Ferrules. Even the turner who enjoys making his own file handles is constantly in need of ferrules. Copper pipe is often the correct size. Silver ferrules are made by joining a band of tie bar with silver solder, filing away the bead, and hammering or rolling the band until it fits. If the ring needs to be expanded by much, it is best to heat the piece to eliminate the effects of strain hardening.

If you favor the harvesting of elephants you

may want to try ivory. The first step is to saw the tusk into slabs the thickness of the ferrule. (Make sure that the grain of the ivory is parallel to the grain of the wood, as this eliminates some of the differential expansion problems.) Cut the slabs into checkerlike discs and hold each piece in a lathe to cut out the center. Figure out a way to save the center so it can at least be used to make a button or an ornamental dot for the end of a walking staff. Ivory is a wonderful material to work with, a lost pleasure akin to the light given off by a whale-oil candle.

Ivory substitutes. In the old days they developed a good ivory substitute used for toilet seats. While this material is prized by guitar makers, it is of little use to the flute maker. My favorite substitute is a material called paper epoxy made of layers of paper bonded by epoxy or some other plastic. This material is surprisingly expensive, has a nice grain, and is available in small quantity through the distributors of knife-making supplies. Naturally they give it their own trade name so as to make it difficult to go to a plastic laminate factory and ask for what you want.

Bone is a wonderful ivory substitute if you don't mind harvesting your fingers. If your digits make it through the band saw, they can look forward to being abraded away in the belt sander. My first inclination was to find a ham bone ring the right size but I found that most bones have a flat side, making this approach impossible. The basic technique is to cut slabs (not rings), sand them flat, then cut discs and bore them out.

English tongue: as an acoustic material wood reflects high and low frequencies about equally, while masonry absorbs more high-frequency tones; if these acoustic properties of wood homes have fostered the sound of English it would explain the way the French, whose homes were predominantly stone and brick, have historically described English, as a "hiss-

A baroque flute by John Ingalls

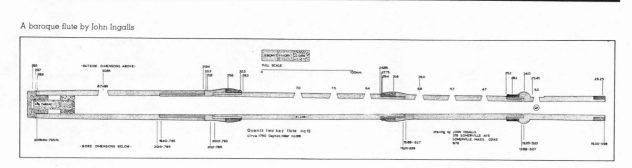

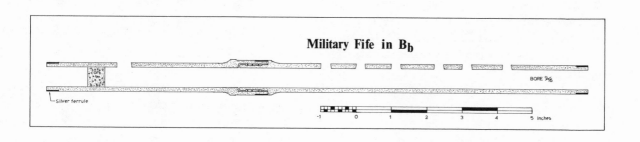

ing language." Wood is a comforting companion. It is good to see the colors and figures of wood. It is good to touch wood, to feel its warmth with your hands. A house without a tree is alone and a room without wood is cold. It is good to smell a wood fire on a cold day and to hear it burning in the stove. It is good

Production Turning

Most of the production flute makers of today seldom use hand tools. Instead, they cut templates to match the outside contour of the flute. The template is placed in a metal lathe and it cuts the exact contours of the flute. This invention changed the flute maker into a glorified key maker. I always considered the outside shape a small step in the building of a flute and have resorted to the key cutter.

Design

When I first became interested in building flutes I sat down and used some elementary physics to calculate the length of the tube and the exact placement of the finger holes. I soon found that the instrument I came up with was totally out of tune. A little reading on the subject pointed out that no one had ever come up with an infallible formula for the exact length of an organ pipe. The flute presents a similar problem but is infinitely more complicated.

There is one way to originate a new flute: trial and error. In other words, one takes a tube, drills holes, and seals them if they are not correct. One can also copy and modify an existing flute.

I would like to close this section by passing on some bits of hard-found information. The first is a blueprint of a Renaissance flute that is easy to build. The other is a trick I found valuable in designing flutes. Quite often I

needed to alter the length of an instrument to achieve a certain pitch but keep all the proportions constant. An easy way to accomplish this is to mark the spacing on a stretched piece of model aircraft rubber band and the proportions will remain constant as the rubber is stretched.

Finishing

As I pointed out earlier, the critical consideration in finishing is to keep the wood of the bore from expanding. Many of the production wooden flute makers have resolved the problem by varnishing or epoxying the bore. Personally, I seldom use this method.

I place the flute in a canister filled with a mixture of boiled linseed oil and turpentine and put it through a series of temperature changes from about 65° to 130°F. These temperature changes tend to draw the mixture into the pores. After soaking for several days I remove the pieces and wipe away every trace of linseed oil. About a week in air allows the oil deep in the pores to polymerize but the surface takes on a leached appearance. The flute is then immersed in the mixture again overnight. This time it is removed but not wiped dry. After eighteen hours the oil begins to form a sticky coating on the surface. At this point the scum is removed with the aid of turpentine and ultra-fine steel wool.

Almond oil is referred to in some early books on flute making. From my experience it does not have enough body for the flute maker. Per-

to smell it fresh cut on your bench and to feel its chips and curls build up around your feet. It is good to make solid, useful things for your home and see them used. It is good to hear music come out of a pale brown guitar and a rich red dulcimer and a deep russet violin. Wood is an old friend.

haps it would be good to maintain the surface. The only flute I ever treated with almond oil cracked.

Conclusion

I have addressed this article to the curious woodworker hoping that he will find information useful in his own particular line of work. I would like to close with a paragraph addressed to the craftsman who is interested in building flutes.

The information included in this article was acquired during a four-year period of my life devoted solely to making flutes. I could pass on more information to you about turning, tuning, and finishing but am not a good enough writer to keep from boring the rest of the world to tears. It is the information I cannot pass on that is most essential in producing an instrument that will sing. In my travels I have found objects resembling flutes. At times I have been able to add and subtract material here and there and make them sing. If you would like to build flutes, do not start with a piece of exotic wood. Pick up a piece of PVC pipe or bamboo and just play with it. After several months you will understand the flute. The rest is just learning to master the materials.

CARVING LETTERS

I asked Fud Benson of the John Stevens Shop in Newport, Rhode Island, to show me how to carve the exquisite letters he produces with apparent ease. No one has a firmer hand or a finer eye for the classic styles of inscription, no one makes it seem easier. Sure, he said, I'll show you just how to do it, but it won't do you any good. He told me a story while he ticked away at a clean majuscule "U" in a dark slate tombstone. "A group of dudes were drifting through the lobby of the old Biltmore in New York. [This is all punctuated by puffs to blow slate dust out of its groove, snorts of satisfaction, and momentary lapses for full concentration on a tight curve.] And Scarne, the card master, was amusing himself by keeping his hands loose sitting there waiting for someone. Shuffling a deck and then cutting to all four aces. Well the dudes were amazed and they were rich. One of them offered him a thousand dollars to show him the trick. Another offered fifteen hundred and the bidding started. 'Wait a minute,' Scarne said, 'I'll tell you how to do it and it won't cost you a cent. Just sit down with some cards for about three years until you know where the aces go when you shuffle them in. Then spend another five years and you can cut down the right number of cards. Simple.' Sure, Bwah, I'll show you how to do it. Show you in five minutes. But you might have to sit down with it for a few years." Puff. Snort.

Fud was as right as any artisan who tells a beginner that the skills of a craft are simple but perfecting them is beyond effort. If you start tomorrow morning and, excluding every other pursuit, work at carving letters for ten years your work will not have the grace and rhythm of the inscriptions that go out the loading door of the Stevens Shop. They are lifetime calligraphers and carvers in the shop; it's in their wrists and shoulders and behind their eyes. Their standards are terribly high. But if you work carefully and with understanding, your third sign will be very attractive. Its inscription will have a grace plastic tack-on letters can't have. Here are the basics.

Start with fairly soft, even-grained wood — basswood or sugar pine is fine — and keep the grain in your mind as you work, never forget it. Work slowly, take out stock with several shallow passes rather than one deep cut. Work toward the lines, not to them, then shave up to them as stealthily as you would steal a mastiff's bone.

Sharpness. The edges on your skews and gouges must be a quantum jump beyond hair-shaving keenness. They must frighten you a little. Fud's tools, already sharp, were brought up to a proper working edge with strokes on a fine India, then a hard Arkansas, then a Belgian honing stone, then on a rouge-dressed leather strap, and finally on plain leather. They were very sharp.

Begin with a skew chisel on the straight cuts. Start near the center with a shallow cut angled in toward the center, reverse the angle for the other side and make a similar pass, freeing a shallow, narrow V groove. Widen it and deepen it in successive passes on both sides toward, but not to, the guide lines. Cut the ends square; the serifs come last. On curves, make an initial curving cut with the skew on the convex side and free the chip from the concave side with a gouge of a sweep that matches the curve at that point (letter carvers need a comprehensive set of gouges; see box). Serifs are extended from the square-cut ends with careful skew chisel work.

All this presupposes the other half of its difficulty, laying down guidelines for beautiful letters. The only help you have here is reference: A good lettering book or the *Speedball Handbook* from an art supply shop will have several *simple* letter faces to copy. Copy them exactly: letters and proportions are fixed in our sense from long association; it takes a special skill to tamper with them successfully. It's as simple and as difficult as that.

The basic tools for carving medium-sized letters of two to six inches, best-quality steel:			
Skew Chisel	1 H 8	Gouge	8 H 5
Skew Chisel	1 H 4	Gouge	8 H 14
Gouge	5 H 12	Gouge	8 H 21
Gouge	6 H 12	X-Acto Knife	
Gouge	6 H 18	Sharpening Stones & Slips	
Gouge	7 H 12	Lettering Book (Speedball Handbook)	

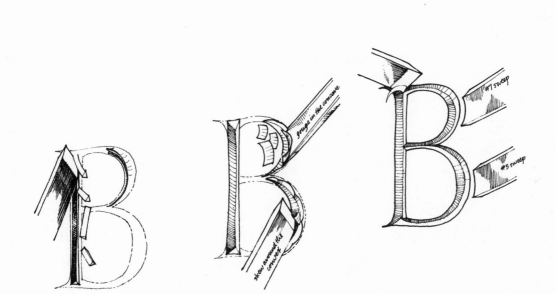